The Nature of the Universe

Books by Fred Hoyle

FRONTIERS OF ASTRONOMY
MAN AND MATERIALISM (WORLD PERSPECTIVES)
THE NATURE OF THE UNIVERSE

Fiction

THE BLACK CLOUD
OSSIAN'S RIDE

The Nature of
THE UNIVERSE
REVISED EDITION

by FRED HOYLE

Plumian Professor of Astronomy, Cambridge University

HARPER & ROW, PUBLISHERS

New York and Evanston

THE NATURE OF THE UNIVERSE, Revised Edition
Copyright © 1950, 1960 by Fred Hoyle
Printed in the United States of America
H-M
All rights in this book are reserved.
No part of the book may be used or reproduced in any manner whatsoever without written permission except in the case of brief quotations embodied in critical articles and reviews. For information address Harper & Row, Publishers, Incorporated, 49 East 33rd Street,
New York 16, N. Y.

Library of Congress catalog card number: 60-13436

CONTENTS

Preface to the First Edition	vii
Preface to the Revised Edition	ix
1. The Earth and Near-by Space	1
2. The Sun and the Stars	27
3. The Origin of the Stars	51
4. The Future and Fate of the Stars	79
5. The Origin of the Earth and the Planets	93
6. The Expanding Universe	106
7. A Personal View	133

PREFACE

To the First Edition

IN PREPARING THESE LECTURES FOR PUBLIcation I have decided to follow the original scripts as closely as possible. There is so much difference between the written and the spoken word that any attempt to cast these talks into a literary mold would have involved a major task of reconstruction. Wherever I have felt additional information to be desirable I have added a note at the end of the book: so if some point in the text should be found puzzling, it may be that a better understanding can be obtained by consulting one of these supplementary notes.

It is not too much to say that but for the constant help and encouragement of my wife the original broadcast lectures would never have been given. I am also deeply indebted to Mr. Peter Laslett, whose sure instinct for the right way "to put an idea across" was in a

PREFACE

large measure responsible for the degree of clarity achieved. Many of the most graphic remarks and phrases were suggested to me by Mr. Laslett.

F. H.

PREFACE

To the Revised Edition

THE OPPORTUNITY HAS BEEN TAKEN TO REarrange and revise the text wherever recent astronomical work makes it necessary to do so. One chapter, that dealing with the origin of the planets, has been entirely rewritten. I have tried not to change the general style and plan of the book.

F. H.

CAMBRIDGE
July, 1959

The Nature of the Universe

1 THE EARTH AND NEAR-BY SPACE

Doubt thou the stars are fire;
Doubt that the sun doth move;
Doubt truth to be a liar;
But never doubt I love.

THUS HAMLET WROTE TO OPHELIA, APPEALing twice to astronomical matters in one verse. For Shakespeare lived in a day when cosmology meant much to the average man. So it is not surprising to find him using cosmological ideas and imagery, one might almost say on every possible occasion. To the Elizabethans a realization of the size of the Earth and of the nature of near-by space was exciting news; it destroyed once and for all the tight little cabbage-patch world in which man had lived throughout the medieval age.

During the following three centuries, although cosmology came to mean more to the mathematicians and astronomers, it had less and less effect on the outlook of people in general. But in recent times—and by recent times I mean the last 50 years or so—it has become increasingly plain that the stage is being set for the next cosmological revolution in our way of thinking. The popularity of the well-known books of Eddington and Jeans was an obvious sign of reawakening interest in the relation of man to the universe as a whole.

As a character says in one of Mr. O'Casey's plays, "What is the stars, what is the stars?" Well then, what is the stars? What is the insides of the stars, and what lies in the depths of space beyond the bright girdle of the Milky Way? Come to that, what is the Universe? Does it end somewhere or does it go on forever or has it a definite age—like you?

This is the sort of thing I shall be dealing with in this book. My general plan, which it may be useful for you to know in advance, is to proceed outward from ourselves. In this chapter we shall begin with the Earth and its immediate surroundings. In the next chapter I shall try to describe what we know about the inner workings of the Sun and similar stars, and to discuss whether energy can be generated on the Earth by processes analogous to those occurring in the Sun. Then in the third chapter we shall move on to matters of wider scope, both in space and time; to deal with the way the stars of the Milky Way are arranged in space, and how they have originated and what their ages are. After that I shall tell you about the origin of the Earth itself. Then in the final chapter we shall be concerned with the Universe as a whole, especially with its expansion and with the question of whether it is being created all the time, while at the end I shall say how, in my opinion, the conclusions of the New Cosmology ought to affect our philosophical and religious outlook.

That is to say, my aim is to put before you the new developments of the last 15 years or so. In some matters this means bringing up ideas that are either current or only a few years old. I shall also have to discuss various issues that are still controversial. So it is important for you to realize that there is no present finality to the discussion of many of the questions that will turn up, and that future work is certain to throw more light on almost every problem.

Before starting the main astronomical discussion there are one or two further points of a general nature that I should like to cover. For a hundred years after the death of Newton, a thorough inquiry into the nature of the Universe was still believed to be impossible. We know now that this belief is not correct, but throughout the eighteenth century there were apparently very good reasons for it. The astronomer is severely handicapped as compared with other scientists. He is forced into a comparatively passive role. He cannot invent his own experiments as the physicist, the chemist, or the biologist can. He cannot travel about the Universe examining the items that interest him. He cannot, for example, skin a star like an onion to see how it works inside.

Yet in spite of these obvious difficulties progress made during the nineteenth century showed that the situation was not so hopeless as it once had seemed. And with the coming of the twentieth century the astronomical scene has changed to a degree that could hardly have been thought possible by the early investigators. Instead of our being, as it were, small boys at the holes in the circus tent struggling to get even an imperfect peep at the great show, we now see that we have really got ringside seats from which the Universe may be observed in all the majesty of its evolution. This transformation has arisen mainly from the work of the American observers, who have exploited with the greatest skill and imagination the large telescopes that can be used in their favorable climate, like the one at Palomar. Just as a blazing bonfire is to a penny candle, so is the observational progress achieved in the last few decades to the work that came before.

But observation with the telescope is not sufficient by itself. Observation tells us, for instance, that while the majority of stars are common or garden specimens like the Sun, others are so brilliant that they act as beacons shining from the depths of space. Still others, although no larger in size than the Earth, are so intensely hot at their surfaces that they emit the short wave-length radiation known as X-rays instead of ordinary light and heat. Finally, at the other extreme of size some stars are so huge that if one of them replaced the Sun, it would fill up most of the space occupied by the solar system; that is to say, the Earth would find itself lying deep inside the gigantic body of the star. But observation alone will not tell us why there are these different sorts of stars or what the connections are between them. Nor does observation tell us of its own accord how the stars have come into being, or what their ages are, or what will ultimately happen to them. To answer these questions and many others like them we have to enter the province of theoretical astronomy.

If I were asked to define theoretical astronomy in one sentence, I should say that it consists of discovering the properties of matter, partly by experiments carried out on the Earth and partly through the detailed observation of near-by space, and in then applying the results to the Universe as a whole. It may reasonably be asked whether this is a valid procedure, and perhaps I had better deal with this question before we go any further. Can we expect the information obtained from one particular small region of space and over one particular small range of time to be applicable at all times throughout all space? For example, did matter in a remote star some 1,000 million years ago behave in basically the same way as matter does on the Earth at the present time? Questions of this kind constitute one of the central issues of Einstein's theory of relativity. For the principle of relativity is simply a statement that our local results do indeed have universal validity. In short, if relativity is correct then our general procedure in theoretical astronomy is guaranteed.

PRINCIPLE OF RELATIVITY

I should like to expand this a little even though the point I want to make may seem rather difficult. The procedure in all branches of physical science, whether in Newton's theory of gravitation, Maxwell's theory of electromagnetism, Einstein's theory of relativity, or the quantum theory, is at root the same. It consists of two steps. The first is to guess by some sort of inspiration a set of mathematical equations. The second step is to associate the symbols used in the equations

THE EARTH AND NEAR-BY SPACE

with measurable physical quantities. Then the connections that are observed to occur between various physical quantities can be obtained theoretically as the solutions of the mathematical equations. This process has two important advantages. It not only makes it possible to condense an enormously complicated mass of experimental information into a few comparatively simple equations, but it also brings out new and previously unsuspected relations between the physical quantities. What Einstein's principle of relativity states is that wherever you are in the Universe, whatever your environment, the same mathematical equations will suffice to describe your observations. I hope that you will agree that this is a very powerful statement.

But it is now high time that we came to the astronomy and there is clearly no better place at which to begin our survey than with the Earth. The surface of the Earth is very nearly a sphere of about 8,000 miles diameter. The inside of the Earth consists of a central core rather more than 3,000 miles in diameter, surrounded by a thick rocky shell that for the most part is more rigid than steel. The essential features of the core, which have been established through the study of earthquake waves, are these: it contains fluid, and its density is substantially higher than the rocks composing the shell. There is fairly general agreement that the fluid consists of molten metals, particularly iron, and that its temperature is about 5,000° C. Partly because the metal is a good conductor of electricity, and partly because it is in turbulent motion, it acts as a gigantic dynamo, generating the magnetic field of the Earth—as Bullard and Elsasser have shown. Whether we use a compass to find our way across the open moors, the desert, or the trackless ocean, we are making practical use of the boiling of metals that goes on ceaselessly inside the core of the Earth.

Rocks at the Earth's surface contain small particles of iron that in some cases became fixed, like tiny magnets, when in the distant past the rocks cooled and solidified. By studying these fossilized magnets it is possible to decide whereabouts the poles of the Earth lay in bygone ages. It is known in this way that the North Pole lay some 500 million years ago in the Pacific Ocean, southeast of the coast of China. Once a complete magnetic survey of the surface rocks of the Earth has been made, it will be possible to settle an old argument: Do the continents have permanent shapes? Was Europe always separated from America?

The Earth extends probing fingers deep into space, for our magnetic field reaches far outward across the space that separates the Earth from the Moon. Recent

THE EARTH AND NEAR-BY SPACE

work with artificial satellites, particularly by Van Allen, has shown that high-speed particles from the Sun become enmeshed in the Earth's magnetic net. The beautiful polar aurora is caused by electrons that become trapped in this way. The electrons race backward and forward from one pole to the other. In fact they move fast enough to make one complete pole-to-pole trip in less than a single second of time.

THE EARTH SEEN FROM OUTSIDE

Mention of satellites means that we must now leave the Earth behind and must move out into the wider world beyond. Once a photograph of the Earth, taken from outside, is available, we shall, in an emotional sense acquire an additional dimension. The common idea of motion is an essentially two-dimensional idea. It concerns only transportation from one place on the surface of the Earth to another. How many of us realize that but for a few miles of atmosphere above our heads we should be frozen as hard as a board every night? Apart from the petty motion of the airplane, motion upward as yet means little to us. But once let the possibility of outward motion become as clear to the average man at a football match as it is to the scientist in his laboratory, once let the sheer isolation of the Earth become plain to every man whatever his nationality or creed, and a new idea as powerful as any in history will be let loose. And I think this not-so-distant development may well be for good, as it must increasingly have the effect of exposing the futility of nationalistic strife. It is in just such a way that the New Cosmology may come to affect the whole organization of society.

Now what does the contemporary astronomer expect such a photograph, a color photograph of the Earth, to look like? Seen through a blue filter there will be brilliant patches where the Sun's light is reflected from clouds and snowfields. The Arctic and Antarctic will on the whole appear brighter than the temperate zones and the tropics. There will be all shades of green, varying from the light green of young crops to the somber darkness of the great northern forests. The deserts will show a dusky red, and the oceans will appear as huge areas that look grimly black, except occasionally they will be illuminated by a blinding flash where conditions allow the Sun's light to be powerfully reflected, much as we sometimes see a brilliant shaft of sunlight reflected from the windows of a distant house. The whole spectacle of the Earth would very likely appear to an interplanetary traveler as more magnificent than any of the other planets.

THE EARTH AND NEAR-BY SPACE

So much then for the Earth. Perhaps we should next take a look at our nearest neighbor, the Moon, which is a ball of rock only about an eightieth as massive as the Earth. The Moon is a satellite; that is to say, it moves along a nearly circular path around the Earth. The time required for it to go once around this path is called the lunar month, which is at present about twenty-seven days. I cannot explain here exactly why it should be so, but owing to the effect of the tides our satellite is getting steadily farther away from us and the lunar month is getting steadily longer. If we go back into the past the converse is true; namely, that the Moon was then nearer to us than it is now. If we worked backward in time as far as the birth of the Earth, which, as I shall show in a later chapter, occurred about 4,500 million years ago, it is probable that the Moon was then very close to the Earth, if indeed the two were not in actual contact. During the aeons that have since elapsed, the tides have not only caused the recession of the Moon, but its rotation has been gradually slowed down. It turned more and more slowly until now it keeps one face permanently toward us. No one has yet seen the opposite side of the Moon.

The Moon has no detectable atmosphere, and its surface is severely pockmarked, looking as if it had been bombarded by a host of large celestial missiles. And this is very likely just what has happened. At one time it was thought that the lunar craters were extinct volcanoes, but for the following reasons this now seems unlikely. Some of the craters are over a hundred miles across, and the big ones show almost the same structure as the small ones. Terrestrial volcanic craters, on the other hand, are only a few miles across and do not show the same uniformity of structure.

Very recently an observation of a disturbance actually taking place on the Moon was claimed by a Russian astronomer, an observation that would be consistent with a volume of gas escaping from the interior of the Moon through the lunar surface rocks. Unfortunately, this work lacks corroboration and is therefore exposed to doubt.

Perhaps the strongest argument in favor of the bombardment theory is that the amount of material in the walls surrounding a crater can actually be estimated, and it turns out to be just the amount required to fill in the hole in the floor of the crater. This has been established through the researches of Ralph Baldwin. But in spite of such a patient clue the bombardment theory has not gained complete currency among astronomers because it is thought that a strong argument can be brought against it. There are large areas of the Moon where no craters can be found. How have all the missiles contrived to miss these areas, whereas in other places the craters are almost overlapping each other? A likely way round this apparent difficulty has been pointed out by my colleague, Gold. The fierce heating of the lunar surface rocks by day and the cooling by night must lead to an alternate contraction and expansion which causes small bits of rock to flake away from the surface. These particles of dust tend to work their way to the lower parts of the Moon, where they have accumulated as gigantic drifts that cover the underlying craters. I think that this new idea may well be correct, because it not only overcomes the old objection, but it also explains those cases where the walls, or a portion of the walls, of a crater stick straight out of an apparently flat plain. These are simply the cases where the drift of dust is not sufficiently deep to cover the craters entirely.

Perhaps I might digress here to deal with an objection to Gold's suggestion. It has been claimed that the slopes on the Moon are mainly of too gentle a gradient to allow particles of rock to drift in the required manner. But this objection assumes that the bits of rock scrape along the ground, whereas according to Gold they are held above the ground by electrical forces. This enables them to float freely down the gentlest of gradients.

Another objection to the bombardment theory is that

many of the missiles must have struck the Moon at highly oblique angles. How comes it then that the craters are all nearly circular in shape? This objection is in error because, just as the damage done by an ordinary bomb is due to the heat produced in the explosion and not to the metal casing of the bomb, so the blast responsible for producing a crater is due to the heat released in the impact of the missile with the lunar surface. This depends on the velocity and size of the missile but not on its direction of flight. The white streaks that are observed to radiate from such a crater as Tycho are probably due to streams of fused material that were shot out in the blast that formed the crater.

I said a few moments ago that no one has as yet seen the other side of the Moon. But we might live to do so. If a chunk of material the size of a mountain should strike the Moon obliquely, there is no doubt that the Moon would appear to us to be set in rotation again, and the unseen side would turn slowly toward us. As you will guess, this is not very likely to happen, but there is nothing impossible about it.

This brings me back to the Earth for a minute. The Earth must have suffered an even greater bombardment than the Moon. But with the exception of one or two recent formations, such as the famous meteor pit in

THE EARTH AND NEAR-BY SPACE

Arizona, the resulting craters on the Earth have been removed through erosion by wind and water. Do missiles still strike the Earth? Well, a piece of rock, probably about the size of a house, hit Siberia in 1908, and the resulting blast is said to have felled trees over a wide area. At any moment another celestial cannonball may hit the Earth anywhere. For all we know, the whole of London may be wiped out during the next five minutes. But there is no occasion for alarm; the odds against it are very large.

Pride of place in our sky belongs to the Sun, not the Moon. The Sun is about 390 times farther away from us than the Moon and is about 300,000 times more massive than the Earth. The Sun shines by its own light, whereas the light from the Moon, and the light that would be seen from the Earth if we could observe it from a distance, is simply reflected sunlight. I don't wish to say much about the Sun now as it forms the main topic of the following chapter. But there is one effect of the Sun—namely its control over the motion of the Earth—that is so important that I feel something must be said about it, even though this means spending the next few paragraphs discussing the history of astronomy.

The direction of the line joining the Sun to the Earth changes from day to day. Starting at a particular time, this line swings round in a plane and comes back to its original position in a year. The direction of the line joining the Earth to any one of the other planets also changes from day to day, but not in such a simple manner. A planet is a body, small in size compared with the Sun and near to us compared with the stars, that shines by reflected sunlight, as the Moon does. In ancient times five planets were known, other than the Earth. They were recognized from their motions in the sky, which have the effect of constantly altering their positions relative to the stars. It was found that they all lie nearly on the plane swept out in the course of the year by the line joining the Earth and the Sun, so that the whole solar system forms a sort of flat platelike structure.

The first attempts to explain all these changes of direction in terms of motions in space were made on the assumption that it is the Earth that is fixed and that the planets as well as the Sun move in orbits round the Earth. Although much ingenuity was displayed by such men as Eudoxus and Ptolemy, the picture they drew was complicated, and as further observational information came to hand it became more so, until it was a thorough mess. As far as we know the first man to perceive that a far simpler description could be achieved by taking the Sun as the center of the system was Aristarchus of Samos,

THE EARTH AND NEAR-BY SPACE

who lived in the third century B.C. He found it possible to explain the observations by supposing that all the planets, the Earth included, move around the Sun in essentially circular orbits of various radii. If sufficiently detailed historical records were available it would be an interesting study in prejudice to see why Aristarchus' views were ignored by his fellow Greeks. At all events they were forgotten until revived by Copernicus nearly 2,000 years later.

The conflict between the Copernican theory and the Roman Catholic Church is well known, especially the part played by Galileo. In recent years there has been an increasing tendency for commentators on this episode to provide an apologia for the attitude of the Church. I am not in sympathy with this, for I regard the statements of Galileo's opponents as exemplifying the dictum of the humorist Josh Billings, which ran something as follows: "It ain't what a man don't know as makes him a fool, but what he does know as ain't so." The case for the Copernican theory is not that it is right or true in some absolute sense, but that it was the only point of view from which progress could have been made at that time. In short, that it had the virtue of simplicity, and this was demonstrated with great cogency and skill by Galileo.

How far Galileo was justified in advocating the

Copernican theory is shown by subsequent events. Astronomy now began to advance in giant strides. Kepler found that the planets do not move exactly in circles but in nearly circular ellipses. The notion of gravitational attraction occurred to many people, including Halley, Wren, Hooke, and Newton in England. By considering the planetary orbits to be circles, a quantitative formulation of gravitation was obtained. But could this formulation explain the ellipses found by Kepler? Only Newton was able to prove this, and in doing so he set the pattern for all subsequent scientific investigations. But Newton did not stop there. He went on to show that many other apparently disconnected observations—the tides, for example—could also be explained by gravitation. His work was on such a colossal scale that it is natural to find the following century and a half given over to the consolidation of his ideas. The detailed explanation, in terms of gravitation, of every feature of the planetary motions became the chief program of mathematicians and astronomers. It was Adams and Leverrier who closed this chapter in the history of science by using the Newtonian theory to predict the existence and the position of a new planet named, subsequent to its observational detection in 1846, Neptune.

From now on astronomy takes a different turn. The motions of the planets cease to be of chief concern. The

THE EARTH AND NEAR-BY SPACE

nature of the Sun and the stars, and latterly nothing less than the whole Universe, provide the subjects to be attacked by scientists. We shall be dealing with these topics in the later chapters of this book. For the present let us take a more detailed look at the planetary system. Suppose we make a plan of how the Sun and the planets are arranged. In our plan let us represent the Sun as a ball six inches in diameter, the sort of thing you could easily hold in one hand. This, by the way, is a reduction in scale of nearly 10,000 million. Now how far away are the planets from our ball? Not a few feet or one or two yards, as many people seem to imagine in their subconscious picture of the solar system, but very much more. Mercury is about 7 yards away, Venus about 13 yards away, the Earth 18 yards away, Mars 27 yards, Jupiter 90 yards, Saturn 170 yards, Uranus about 350 yards, Neptune 540 yards, and Pluto 710 yards. On this scale the Earth is represented by a speck of dust and the nearest stars are about 2,000 miles away. You may wonder how the various distances in the solar system have been established. In principle the methods used are only a matter of elementary trigonometry, but in practice very intricate measurements are necessary and even then extensive calculations have to be made in order to extract the required information.

The wide spacing of the planets is even more remarkable than our plan would suggest because the biggest planets are the ones lying at great distances from the Sun. Venus is slightly smaller and Mercury and Mars appreciably smaller than the Earth. Jupiter, on the other hand, is more than 300 times as massive as the Earth, but even so Jupiter is still much smaller than the Sun. Indeed, if the Sun were suitably divided up it would make more than 1,000 bodies of the size of Jupiter. Saturn is about 95 times more massive than the Earth, Uranus about 15 times, and Neptune 17 times. The exact size of Pluto is not known, but it is certain that Pluto is small like Mercury, Mars, and Venus. The four large planets, Jupiter, Saturn, Uranus, and Neptune, are often referred to as the great planets, and they lie at distances from the Sun that, on our plan, exceed 90 yards. I have been insistent on getting the scale of the planetary system right, because an appreciation of the wide spacing will be of great help when later we come to consider how the planets originated.

Let us leave the solar system for the present by taking a more detailed look at some of its members. Venus might be called the twin sister of the Earth, and therefore naturally claims our attention first. Venus is not very aptly named, for she modestly hides her surface by a perpetual bank of white cloud. The nature of this cloud is a bit of a mystery. At first sight it might be thought to consist of water drops, as the clouds on Earth do. But it seems rather unlikely that this idea is correct, otherwise the presence of water vapor would surely have been detected. So far the only gas found in the atmosphere of Venus is carbon dioxide, which is present in enormous quantities. All attempts to detect substances with cloud-forming properties similar to water have failed.

There is another mystery about Venus. According to the observational astronomers it takes her more than 20 days to rotate on her axis. As it seems probable that all the planets at their birth had rotation periods of about ten hours, it is an interesting question as to what process has slowed down the rotation of Venus to such a marked extent. I cannot enter now into any detail beyond saying that one possibility is that the Sun exerts a huge tidal influence on an ocean of some liquid that may cover much of the surface of Venus, and that the only other possibility known to me is a suggestion by Lyttleton of Cambridge, that Mercury may once have been a satellite of Venus.

The most interesting feature connected with the planet Mercury concerns its orbital motion around the Sun. The ellipse in which Mercury moves is more flattened than the orbit of any other planet but Pluto, and the strange discovery made about 75 years ago by the famous

French mathematician, Leverrier, was that, quite apart from disturbances due to the known planets, the longest axis of the ellipse is slowly turning in space. At first this was thought to be an effect due to a new planet moving in an orbit still nearer to the Sun. So sure were astronomers of the existence of this hypothetical planet, especially after their experience concerning the discovery of Neptune, that they had the name Vulcan ready to be attached to it as soon as its existence was confirmed by observation. But it is curious how rarely in science the same tactics can be repeated with success. Whereas Neptune had been detected almost immediately once the predictions of Adams and Leverrier were made known, the observers now searched in vain for Vulcan. It is also curious that Leverrier's failure to find a new planet in his second attempt should turn out to be of far greater scientific importance than success would have been. For success would only have added a small body to the solar system, whereas failure meant that at long last a flaw in the Newtonian theory of gravitation had been found. About 40 years later the resolution of this issue was to form one of the cornerstones of Einstein's general theory of relativity.

Now let us move outward from the Earth in a direction away from the Sun. What can we expect to find on the first planet, Mars? Much very interesting information has been obtained by G. P. Kuiper, who has used an ingenious new technique to show that the white polar caps of Mars are indeed composed of snow. The atmosphere contains more carbon dioxide than that of the Earth, and there may also be nitrogen. The thin clouds responsible for the haze so troublesome to the observer are probably composed of ice crystals. No oxygen can be detected, though free oxygen may once have existed on Mars in appreciable quantities. There are dark markings on the surface of Mars that seem to vary from time to time. The changes in these markings may well be due to the growth and decay of plants similar to the rock lichens found on the Earth. It is not surprising that plant life on Mars should be confined to such a low form, for lichens require little moisture and they can survive at lower temperatures than the more usual terrestrial plants. Both these properties would be important requirements on Mars. Are there also other forms of life? I do not think any other answer can be given to this question but to say that we must wait and see.

Beyond Mars lie the great planets, Jupiter, Saturn, Uranus, and Neptune. They are well furnished with satellites; Jupiter has at least twelve, Saturn nine, Uranus five, and Neptune two. Their atmospheres are about as different from that of the Earth as you could imagine, being composed of methane, ammonia, and probably hydrogen and helium. An old fallacy, exposed about 20 years ago by Jeffreys, was that on account of their large sizes the surfaces of the great planets must be hot. Observation confirms Jeffreys' argument and shows that their temperatures are indeed lower than about -150° C. Below a thin fringe of atmosphere, it is probable that in each of these planets there are thick shells of helium and hydrogen overlying an inner ball of ice and rock.

The hydrogen is not gaseous like the ordinary hydrogen of a chemist's laboratory, but is probably liquid, and it acts like a metal. This liquid metallic hydrogen is also probably boiling like the metals in the core of the Earth, and as in the Earth it probably generates a magnetic field. Tremendous electrical storms are continually taking place in the atmosphere of Jupiter, storms that emit intense radio waves. These outbursts are on a far vaster scale than the storms of the terrestrial atmosphere and they seem to be always occurring at particular places, instead of wandering about from one point to another as storms on the Earth do.

There are many questions that we should like to be able to answer about the great planets. What is the red spot that changes so markedly from time to time in the atmosphere of Jupiter? Is it a solid of some sort floating in a sea of gas? What are the particles that form the rings of Saturn? Are they indeed small ice crystals? Are the five inner satellites of Saturn really gigantic snowballs? It would be possible to go on for a long time asking this sort of question, and, interesting as their discussion might be, there would still remain the larger problem of the origin of the planets themselves. But as this issue arises again in a later chapter, I will make an end by asking whether the fragment of space that we have considered here has anything exceptional to distinguish it from all other parts of the Universe. Is this procession of night and day, this movement of the Earth and planets around the warming Sun, something really special, or are there lots of places where similar systems occur? When you look at the heavens, how many of the stars you see have planets encircling them and on how many of these planets might living creatures look out on a very similar scene?

To give a numerical estimate I would say that rather more than 1,000 million stars in the Milky Way possess planets on which you might live without undue discomfort. If you were suddenly transported to one of them, you would no doubt find many important changes, but the changes would not be as remarkable as the similarities. And if you had been brought up from birth on another planet moving around another star you would feel the same sort of emotional attachment to it as we feel for the Earth. You would look out on the Milky Way and wonder, as we do, what proportion of the stars also have attendant planets with living creatures on them. One of the comparatively insignificant stars that you would see would be the Sun. But, even with a powerful telescope, you would not see the Earth or any of the planets of the Sun's system, for they are far too faint to be seen at immense distances. In short, you would have to infer our existence just as we have to infer yours.

2 THE SUN AND THE STARS

THE MAIN PURPOSE OF THIS CHAPTER IS TO tell you what the astrophysicists have discovered recently about the inner workings of the Sun. And this will bring up their answers to a number of age-old cosmological questions. What is the Sun made of? How hot is it? Is it simply hot on its surface, or is the whole body hot, inside and outside? These are some of the things that puzzle people. Much more important is this one. What is the source of the Sun's energy? Is it growing hotter, or colder? How long will it continue to radiate light and heat at just the rate required by living creatures on the Earth? Is it getting smaller and smaller, or will it stay the same size—or even perhaps get bigger? Some of these questions, I might warn you, will take us forward into the remote future, perhaps to a time 5,000 million years hence.

After all this, there is another class of question, not of such wide cosmic importance but of urgent practical interest, that we must also consider. For in the study of the Sun's light and heat astronomy comes in contact with everyday affairs. Not only is sunlight a necessity for the support of life on the Earth, but it is also the ultimate source of all the energy at present used in industry. The power produced by coal and oil represents sunlight that was stored in trees and plants thousands of centuries ago. Even hydroelectric power really comes from the Sun, for it is the Sun's heat the sucks water from the oceans into the atmosphere. But falling water is not a big source of energy, and it is well known that our coal and oil supplies cannot last for more than a few centuries. So it looks as if our power may finally give out, and with it the whole of our present civilization. Moreover, we need more energy, great quantities of it, if we are to go on developing at the rate we are getting used to. How are we to find a new supply of energy? Should we start growing plants with the object of trapping the Sun's light, or should we build a whole lot of miniature suns of our own? This can be done, as you know, by disintegrating uranium in an atomic pile. This will bring me to our newest, our most anxious, fear. It has been maintained by some people that an atomic explosion might fire off a nuclear chain reaction that would blow up the whole Earth. Whether this is so or not must form a part of our cosmology.

THE SUN AND THE STARS

First, then, a few general remarks about the Sun. It is the nearest of the stars—a hot self-luminous globe. Though only a star of moderate size, the Sun is enormously greater than the Earth and the other planets. It contains about 1,000 times as much material as Jupiter, the largest planet, and over 300,000 times as much as the Earth. Its gravitational attraction controls the motions of the planets, and its rays supply the energy that maintains nearly every form of activity on the surface of the Earth. There are some exceptions to this general rule: for instance, the upheaval of mountain ranges and the outbursts of volcanoes.

You might like to ask why the Sun is able to supply its own light, heat, and energy, whereas the Earth and the other planets only shine feebly with the aid of borrowed light? Strange as it may seem, it is best to start this problem by considering the interior of the Earth. Owing to the weight of the overlying rocks, material near the center of the Earth is subjected to enormous pressure. Indeed, in the deep interior the pressures amount to nearly 100 million pounds per square inch. It is the same inside the other planets, and in those that are larger than the Earth the pressures developed are even greater. It may surprise you that the ordinary solids and liquids of common experience should be able to withstand such terrific forces without giving way.

But if we apply this argument to the Sun, we get a different answer. It can be established that, in order to withstand the weight of the overlying layers, the pressure at the center of the Sun must be nearly 100,000 times greater than the already tremendous values occurring within the Earth. Ordinary solids and liquids certainly cannot stand up to compressional forces as great as that. If the sun were constituted like the Earth, it would collapse visibly before our eyes under the inexorable power of its own gravitational field. How then can the astrophysicist explain why the Sun does not collapse, and also why it has remained pretty much its present size, as the geologists have shown, over at least the last 500 million years? There is only one possibility. The material inside the Sun must be hot, very hot, by our standards. By calculation we have discovered that near the Sun's center the temperature must be in the neighborhood of 15,000,000° C. This may be compared with the temperatures in an electric furnace, which are less than 3,000° C., or even with the surface of the Sun where the temperature is only 6,000° C. Here then we can tick off the answers to one of our original questions; namely, how hot is it inside the Sun? It is about 15,000,000° C., and it is very, very much hotter inside than it is at the surface.

THE SUN AND THE STARS

SOLAR RADIATION

It is difficult to appreciate what a temperature of 15,000,000° C. means. If the solar surface and not the center were as hot as this, the radiation emitted into space would be so great that the whole Earth would be vaporized within a few minutes. Indeed, this is just what would happen if some cosmic giant were to peel off the outer layers of the Sun like skinning an orange, for the tremendously hot inner regions would then be exposed. Fortunately, no such circumstance is possible, and the outer layers of the Sun provide a sort of blanket that protects us from its inner fires. Yet in spite of these blanketing layers some energy must leak through from the Sun's center to its outer regions, and this leakage is of just the right amount to compensate for the radiation emitted by the surface into surrounding space. For if the amount leaking through were greater than the amount radiated, the surface would simply warm up until an exact balance was reached. The situation has some similarities with what happens if you heat a long metal bar at one end. Heat travels from the hotter end to the cooler end. But this analogy is not perfect; analogies never are. Heat is carried along a metal bar by conduction, whereas in the Sun the outward leak of energy is carried by radiation. The radiation changes its character as it works its way outward. At the surface it is ordinary light and heat, but in the central regions it takes the form of the very short wave-length radiation known as X-rays.

We now reach an important point. The rate at which radiation leaks through from the central regions and thence into outer space can be calculated—that is to say, the brightness of the Sun can be predicted theoretically. The result of the calculation depends most strongly on the amount of material present in the Sun. If, for instance, the amount were increased tenfold, the brightness would increase about a thousandfold. Not even the most enthusiastic sunbather would welcome this change, for it would cause the whole body of the Earth to melt, and the rocks would bubble merrily. Then again the Sun's brightness depends on the chemical composition of its material, and also on its size. The Sun would become dimmer if it were expanded and more brilliant if it were contracted.

The first calculations along these lines were made by Eddington. In his remarkable book *The Internal Constitution of the Stars,* he worked out a theoretical value for the brightness of the Sun, using as the ingredients of the calculation the quantity of material in the Sun and its known size. The theory gave a brightness nearly a hundred times too large; that is, a hundred times greater than it is known to be by observation—ordinary observation by telescope. But this was not as bad as it sounds, because Eddington had to make a guess at the chemical composition of the solar material. His first guess was that the material is predominantly composed of iron and other elements of what is called high atomic weight. The important feature of this guess was that no appreciable quantities of hydrogen and helium were thought to be present.

By about 1930, Eddington, however, had come round to the view that his original idea of a Sun made of iron was to blame for the trouble. It was found that the presence of appreciable quantities of hydrogen—the element with the lowest atomic weight of all—would make a very big difference in the theoretical result. To bring theory into line with observation, the Sun had to contain either about 35 per cent hydrogen and no helium, or about 98 per cent of hydrogen and helium together—most of the mixture being hydrogen. Now astronomers were effectively unanimous in preferring the 35 per cent alternative, even though H. N. Russell had shown that hydrogen is overwhelmingly predominant in the atmospheres of many stars. Here you must allow me a slight digression, for you see now the working of prejudice. Previous opinion had been that the Sun contained next to no hydrogen. When Eddington's work upset this notion it was decided to accept the lesser of two evils and the 35 per cent possibility was accordingly adopted. And this view has persisted until quite recently. A proper appreciation of the general cosmic abundance of the various chemical elements is, as we shall see, one of the most recent cosmological developments.

Here now is a crucial turning point in our argument. We have seen that the interior has to be very hot indeed to prevent the Sun's collapsing catastrophically. We have also seen how the rate at which the surface radiates energy into surrounding space can be calculated. I have mentioned the various items of information that constitute the basis of the calculation, and I hope that you will have noticed that at no point did I introduce the idea that the Sun actually generates energy by nuclear transmutations taking place in its interior—the sort of thing that goes on in atomic piles. Does this mean that the brightness of the Sun is independent of any such energy's being produced or not? The answer to this is, yes. If the size of the Sun and the quantity and composition of the material it contains are all known, then its brightness is a fixed quantity, quite regardless of whether or not energy production occurs in the interior. This result may strike you as very surprising. Eddington's contemporaries certainly found it so.

THE SUN AND THE STARS

But is nuclear transmutation taking place in the Sun nevertheless? The best way for us to make further progress in this problem is by asking how Eddington was able to deduce that energy generation must indeed be taking place inside the Sun, and at such a rate as to compensate exactly for what is being radiated into surrounding space. Let us suppose, by some magic, that we remove the sources of the solar energy. There will be no immediate change in the Sun's brightness. But, as you will realize, the Sun cannot go on losing energy indefinitely without there being some important changes in its internal structure. What would the changes be? I suppose the natural answer would be to say that the Sun would cool off. But this is wrong. For, as we have seen, the inner regions could not then support the weight of the overlying layers and there would be a complete collapse of the whole body. So a cooling-off process would not be a stable one. The loss of radiant energy from the surface would lead to a very slow contraction of the whole of the Sun, and, paradoxical as it may sound, this compression would actually heat up the material. Eddington's method of determining the brightness remains valid and shows that so far from cooling off, the Sun would actually grow steadily brighter as it contracted. Calculation shows further that the reduction of the diameter of the Sun would be about a hundred yards every year. At first sight this appears to be very little—it would certainly lead to no noticeable effect, even with sensitive instruments, over the whole course of recorded history. But this is only a way of saying that the period of recorded history is extremely short. Over periods of time that are commonplace to the geologists the Sun would change a very great deal.

GENERATION OF ENERGY

If we put this argument in a slightly different form, we can immediately reach our conclusion. For if throughout the geological ages some source of internal energy had not just compensated for the radiation that was being lost at the solar surface, the Sun would necessarily have shrunk by now to a tiny body. In short, it would have become much less than it is observed to be at present.

But the inference that there must be energy generation inside the Sun does not settle our difficulties. We have still to find out exactly how the energy is produced. Ordinary chemical sources are hopelessly inadequate. If, for instance, the Sun were made out of a mixture of oxygen and the best quality coal, the coal would be reduced entirely to ashes in only two or three thousand years. Nor is the natural radioactivity of uranium, such as occurs in the rocks that compose the Earth's crust, sufficient to run the solar engine. Some new source depending on atomic transmutation is necessary. This requirement first made it clear to scientists that it must be possible to find nuclear processes that are very powerful sources of energy. Here, as with so many other important ideas in physics, the lead was supplied by astronomy.

How, then, is energy generated in the Sun? Two suggestions as to this were made by Jeans. One was that the Sun might contain superradioactive material not present on the Earth, and the other that matter might even be annihilated under the physical conditions occurring in the solar interior. For various reasons that it would take me too long to describe, neither of these ideas has passed into current astrophysics. The solution of the problem lay along different lines, and, at the risk of being a little technical, I should like to go over the main developments as they occurred.

Let us transfer the scene to the interior of the atom instead of the interior of the Sun. The chemical elements are classified according to the particles contained in their central nuclei. At the lower end of the list of atoms found in nature is ordinary hydrogen with a nucleus containing one particle—a proton—while at the upper end is the commonest form of uranium, which has a complex nucleus made up of 92 protons and 146 neutrons. I hope you are familiar enough with these terms not to let them worry you.

Measurements made by Aston of Cambridge in the early nineteen-twenties showed that the best way of getting energy out of the elements at the upper end of the list is to break up the nucleus, preferably into two pieces of about the same size. As you are probably aware, the only naturally occurring elements for which this has so far been found practicable are uranium and thorium. Exactly the opposite situation occurs for nuclei containing less than about 50 particles. These have to be added together for energy to be obtained. Many such building-up processes are possible, but only one is of interest in relation to the Sun. Helium is next to hydrogen at the lower end of the scale of atomic weights. If four protons could be combined so as to form an alpha particle, as the nucleus of helium is usually called, a large amount of energy would be set free. Remembering that Eddington's work showed that the sun must contain at least 35 per cent hydrogen, we are naturally led to ask the question: is the conversion of hydrogen into helium the process that supplies the solar energy generation?

THE SUN AND THE STARS

An important start toward answering this question was made in the early nineteen-thirties by Atkinson and Houtermans, who showed that nuclear transformation processes do indeed occur in the solar interior at roughly the required rate. The next step was taken in 1938 by Gamow and Teller, whose work may be described as bringing the ideas of Atkinson and Houtermans into line with the rapidly developing science of nuclear physics. But so far no one had earmarked the actual processes that supply the Sun's energy. This link in the chain was left to H. A. Bethe and C. L. Critchfield, who described the relevant processes in 1939 in much the same terms that we should apply today.

It was at this stage that my colleague, R. A. Lyttleton, and I first became interested in the problem of the structure of the Sun. It seemed to us that Bethe's work, if it were put into the calculations at the beginning instead of at the end, should lead to a considerable improvement in the whole method of investigation. Many of the obscurities and difficulties of the work of the 1920's and '30's were due at root to the use of the observed size of the Sun as a datum of the calculations. So long as the mode of energy generation was unknown, this was a necessary procedure, but once the nuclear processes occurring in the Sun were understood, it was possible to put the whole problem in a much more direct and challenging form. Given only the amount and the composition of the solar material, is it possible to decide purely by calculation both the brightness of the Sun and what its size must be? Lyttleton and I found that this could indeed be done, and we were able to show that the results of the mathematical theory agree with observation to an accuracy of a few per cent. It is a strange thought that in some ways we know more about the inside of the Sun than we do about the motion of boiling water in a kettle.

This is not the end of the story. The next step leads us away from the Sun to other stars. Eddington, right from the outset, was not slow to see his theory also applied to the stars in general. His comparison of theory with observation for a group of about 20 stars was at first regarded as very encouraging, but as time went on certain discrepancies became more and more manifest. These discrepancies persisted until about ten years ago, when it was realized that they can be completely resolved by a change in our view as to the chemical composition of the material composing the Sun and the stars. You will remember that in Eddington's work, consistency between theory and observation could be obtained if the Sun contained either 35 per cent of hydrogen and no helium, or about 98 per cent of hydrogen and helium together. The only step that was necessary to overcome the discrepancies I have just mentioned was to adopt the larger percentage. To sum up the most recent conclusions, a normal star at the time of its birth consists of about 2 per cent oxygen, nitrogen, carbon, and heavy elements such as iron, perhaps 25 per cent helium, and the rest hydrogen. This answers one of our original questions: what is the Sun made of?

At this stage we may notice an important point relating to the origin of the planets. If the weight of hydrogen in the Sun is about a hundred times greater than the combined weight of such elements as silicon, oxygen, iron, and magnesium, it is clear that the composition of the Sun is very different from that of the Earth, where the contribution of hydrogen to the mass is less than $\frac{1}{100}$ per cent. In short, there is about a millionfold difference between the hydrogen content of the Sun and the Earth, and this must be taken into account when in a later talk we come to consider the process that led to the formation of the planets.

By now we have implicitly answered a number of our original questions. How long will the Sun continue to radiate light and heat at just the rate required by living creatures on the Earth? Calculation shows that the supply of hydrogen in the Sun will last for about 5,000 million years. This does not quite answer our question, because after about 2,000 million years the Sun will be getting too warm for our comfort. In other words, as more and more hydrogen gets converted into helium, the Sun will get hotter and hotter. This is another of those results that go the opposite way from what you might naturally expect. By the time the Sun has used about a third of its present store of hydrogen the climate, even at the poles of the Earth, will be getting too hot for any forms of life that at present inhabit it. At a still later stage, the Sun will become so hot that the oceans will boil and life will become extinct. So life will perish in the solar system as a whole, for the same considerations also affect Mars, not because the Sun becomes too feeble, but because we shall be roasted.

I think this answers all our questions concerning the Sun except this one—perhaps a minor one for those whose interest in cosmology is not professional: Is the Sun going to change its size? Is it going to swell or to shrink? To deal with this, we must return to the stars. So far I have spoken as if all stars can be fitted into the scheme I have described. This is not so. Those that do have a special name: they are called main-sequence stars. There are several groups of nonconformers, and one of them, the red giants, I want now to consider briefly. The red giants are also of interest partly because the problem they raise is one of the most recent to receive solution, and partly because, as I shall describe in a later chapter, they serve as a clock whereby we can determine the ages of the stars. I am rather sorry that I cannot bring in all the varieties of nonconformers, as they have such interesting names—the red giants, the white dwarfs, the blue giants, the black dwarfs, the subgiants, and the supergiants.

SIZE AND DESTINY OF THE RED GIANTS

The red giants are normal enough as far as the amounts of energy radiated from their surfaces are concerned. Where they differ from main-sequence stars is in being of much greater size, and of very much smaller average density. They really are big. It was these stars that I had in mind when I said that if the Sun were replaced by one of them it would fill up most of the space inside the Earth's orbit and that the Earth might even find itself lying deep inside the gigantic body of the star.

The line of argument I have described, starting from Eddington's calculations and ending with the discovery of the nuclear processes that take place within the stars, fails completely for the red giants. The reason is very simple. If Eddington's calculation is supplied to them, it gives central temperatures that are far too low for the nuclear reactions to work efficiently.

I have spoken of a different procedure from Eddington's, whereby the nuclear reactions are put into the theory at the beginning and in which the size of the star was an outcome of the calculations. Lyttleton and I were naturally interested to see whether stars as huge as the red giants could be represented by the theory. We soon realized that this was impossible so long as the chemical composition was taken to be uniform throughout the star, as it was in Eddington's theory. Now the idea that every star has a uniform composition had been accepted by almost every author from the earliest attempts on the problems of stellar structure. Was there any reason for believing it to be a correct hypothesis?

In every normal star, hydrogen is being converted into helium inside the central regions, so that appreciable non-uniformity of composition must arise in any star that has already consumed a large fraction of its original supply of hydrogen, unless some process mixes up the helium with the material in the rest of the star. Given certain things in a star, a sufficiently fast rotation, and perhaps a certain sort of magnetic field, such uniform mixing might go on. When adequate mixing occurs, the star re-

THE SUN AND THE STARS

mains of the main-sequence type; that is, its radius remains comparatively small. But when there is only partial mixing or no mixing at all, it transpires that the star must swell as the hydrogen is consumed. Indeed, when 20 to 30 per cent of the inner hydrogen has been converted to helium, calculation shows that the distension of the star becomes exactly of the order observed in the red giants. The importance of all this to the astrophysicist is that when we observe a star with a greatly distended bulk, we know that this star has had sufficient time since it was born to consume a good deal of its initial supply of hydrogen. As we shall see later, this result enables us to work out the ages of the stars with considerable accuracy.

In small stars such as the Sun, that have not yet lived long enough to burn up much of their hydrogen, there can be no great degree of swelling. Even so, it seems that there has already been a slight expansion of the Sun, and this gives us the probable answer to our final astrophysical question. As the Sun steadily grills the Earth it will swell, at first slowly and then with increasing rapidity, until it swallows the inner planets one by one: first Mercury, then Venus, then possibly the Earth. This particular part of the New Cosmology seems to fit in well with medieval ideas about hell.

THE NATURE OF THE UNIVERSE

ATOMIC ENERGY

My last points in this chapter are about terrestrial sources of energy and the possibility of blowing up the Earth.

Is it possible to produce an atomic explosion that starts a chain reaction in the Earth itself? In particular, could some reaction fire off the hydrogen that is present in water, especially in the water of the oceans? If all the hydrogen in the oceans were suddenly converted into helium, the Earth would be vaporized practically instantaneously. The blaze of radiation produced would be as large as the total emission from the Sun added up throughout a whole year, and if there is life on Mars, it would rapidly be reduced to ashes. If you ever mention the end of the world, that is the sort of end you should have in mind.

A high temperature is necessary before hydrogen is affected by nuclear reactions. The highest temperature that can be produced on the Earth occurs in a volume a few centimeters across for a time of about a ten-millionth of a second during the explosion of a uranium or plutonium bomb. This temperature is about 150,000,000° C., which is about ten times greater than the temperature at the solar center. The first question is whether such a bomb exploded under water would act as a detonator to the hydrogen in the water.

I well remember an anxious calculation made in the autumn of 1945. It was with great relief that I found that the temperature produced by the bomb lasts for too short a time for this to be possible. Otherwise an ultimate disaster would presumably have happened before this.

THE HYDROGEN BOMB

But before we leave this subject we must also consider the possibility of the underwater explosion of bombs more violent than the uranium bomb. These considerations have particular relevance to the hydrogen bomb. The idea of a hydrogen bomb is to produce an extremely rapid conversion of hydrogen into helium: to do what the Sun does, but to do it quickly. For this, two conditions are necessary. One is a high temperature, and this is achieved by using a uranium bomb as a detonator. The other necessity is to find a far faster reaction than the main processes that occur in the Sun and the stars. At first sight it looks as though this is an impossibility, because any process that can be used on the Earth can also occur in the Sun. But this overlooks a crucial point. The fastest reacting substances are so extremely rare in the material of the stars that they are not important in astrophysics. In fact these substances have already been used up inside the Sun. But they still occur on the Earth and are available to mankind—for good or ill.

The most powerfully reacting substance is a special form of hydrogen known as tritium, which differs from ordinary hydrogen—the form of hydrogen I have been considering so far—in its central nucleus. The nucleus of ordinary hydrogen consists of one particle—a proton—whereas tritium, as its name implies, has a nucleus containing three particles—a proton and two neutrons. The most powerful reaction is obtained by combining tritium with a form of hydrogen known as deuterium, the central nucleus of which contains two particles—a proton and a neutron.

But to return to our main topic: can the hydrogen bomb explode the oceans? Even with the most violent reaction, this is impossible. The importance of the hydrogen bomb from a military point of view is that it can be made as large as practical questions allow, whereas the uranium bomb is severely limited in size. So a hydrogen bomb of extremely great explosive power can be made if the necessary quantity of tritium can be produced in some way. But it is not the total quantity of energy released by the bomb that decides whether the oceans will explode.

THE SUN AND THE STARS

The crucial quantity is the temperature produced, and curiously enough this must be nearly the same in the hydrogen bomb as in the uranium bomb. So we may conclude that although mankind may engage in foolish personal destruction, the Earth itself is safe.

I come now to the final question: Is it possible to generate energy from hydrogen in a peaceful controlled way here on the Earth? From the common form of hydrogen —the sort with a nucleus consisting of a single proton— no. Nor is the fast bomb reaction between tritium and deuterium very useful, because there is difficulty in securing adequate supplies of tritium. The favored possibility is a reaction involving deuterium alone. If this could be controlled successfully the potentialities would be enormous. Every ton of water could be made to yield as much energy as a thousand tons or more of coal.

A very high temperature is necessary to make the reaction "go." Extremely hot deuterium gas must be held imprisoned in space in such a way that it never touches the walls of the containing vessel, otherwise heat is lost and the reaction stops. Complicated magnetic fields are being used to hold the gas in the hope that it will stay inside a sort of magnetic prison. So far it refuses to do so. Eventual temperatures as high as 300,000,000° will be needed. The temperatures achieved in ZETA, and in other laboratory devices, still lie only in the range from 3 to 5,000,000°. Obviously the day when we shall be able to do what the Sun achieves so easily—convert hydrogen into helium—is still far off. Nevertheless, the problem is of surpassing importance, for success would place the human species beyond the danger of any energy shortage—even if we look a million years or more into the future. We can plainly afford to take many years over the solution of this tremendous problem.

3 THE ORIGIN OF THE STARS

SEVERAL SCENES IN NATURE ARE OF OVERpowering splendor. Sunrise or sunset, especially when seen in the high mountains, is one of them. So also is the sight of the stars in the heavens. The stars are best seen as a spectacle, not from everyday surroundings where trees and buildings, to say nothing of street lighting, distract the attention too much, but from a steep mountainside on a clear night, or from a ship at sea. Then the vault of heaven appears incredibly large and seems to be covered by an uncountable number of fiery points of light.

Surprisingly, the number of stars that actually can be seen at any time with the unaided eye is only a little over two thousand. These stars all belong to what is usually called the Galaxy, and it is about the Galaxy, our Galaxy, that I shall now be concerned. Ours is not the only galaxy in the Universe. It is possible to pick up faint traces of other galaxies when you look at the night sky. But all the stars you can see clearly belong to it and their number increases very rapidly when you do not have to depend on the naked eye. With even a small telescope you can distinguish about a million stars; with large telescopes, like the ones at Mount Wilson, the number rises to well over 100 million—all within this one galaxy.

A glance at the sky will show you that the stars are not uniformly distributed over it. There is a bright band of light, that people call the Milky Way, in which particularly large numbers are concentrated. The stars take on this appearance because the Galaxy is shaped like a disk. When you look at the Milky Way, you are looking along the disk, and so you see a large number of distant stars. But when you look at other parts of the sky, you are looking out of the disk, and you then see only a comparatively few stars—these are just the ones that happen to lie close to us. It is because of their nearness that so many of these stars appear bright.

Now I should like to give you some idea of the size of the Milky Way, and of the distances between the stars. Ordinary units, such as the mile, are not much good for this purpose. As you know, in many astronomical discussions it is best to use light as a measure of distance. It takes light rather more than a second to travel from the Moon to the Earth, for instance, and we can speak of the distance of the Moon as being rather more than one light second. It takes light about eight minutes to travel to

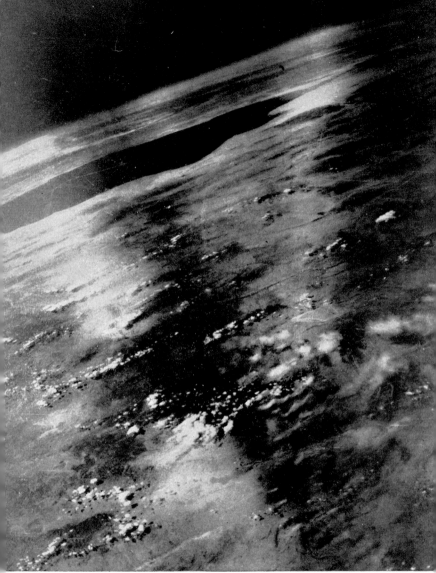

Naval Research Laboratory photograph

PHOTOGRAPH TAKEN BY ROCKET-CAMERA

This shows the curvature of the earth seen from an altitude of 100 miles. The dark patch is the Gulf of California. The width of the area photographed is approximately 250 miles.

Mount Wilson Observatory

CRATERS ON THE MOON

Portion of the moon at last Quarter, including the Apennines, the Alps and Mare Imbrium.

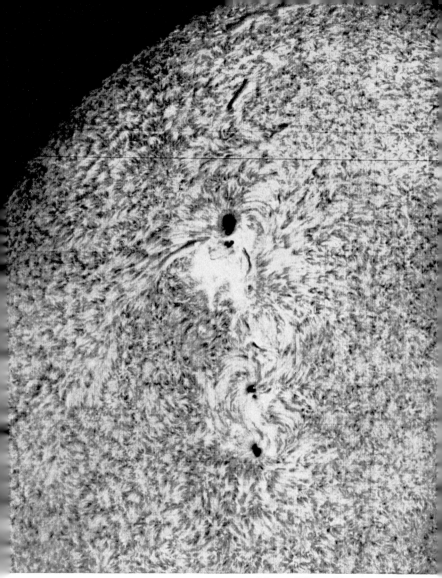

Mount Wilson Observatory

THE SUN'S ATMOSPHERE OF HYDROGEN
This photograph was obtained with the aid of the spectroheliograph, which isolates a faint line given out by hydrogen.

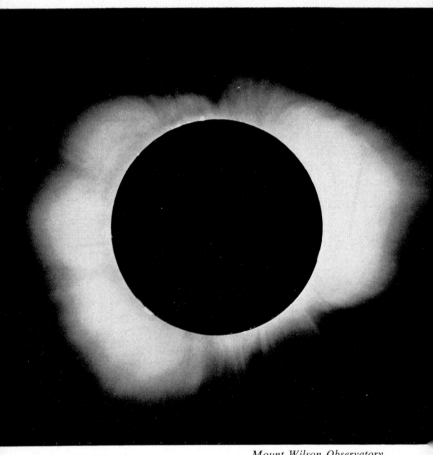

Mount Wilson Observatory

Solar Corona, taken at Green River, Wyoming, June 8, 1918.

Yerkes Observatory

Mars, enlargement of best exposure of September 28, 1909.

Yerkes Observatory

Jupiter, photographed in ultra-violet light by Ross with the 60-inch reflector at Mt. Wilson, August 20, 1927.

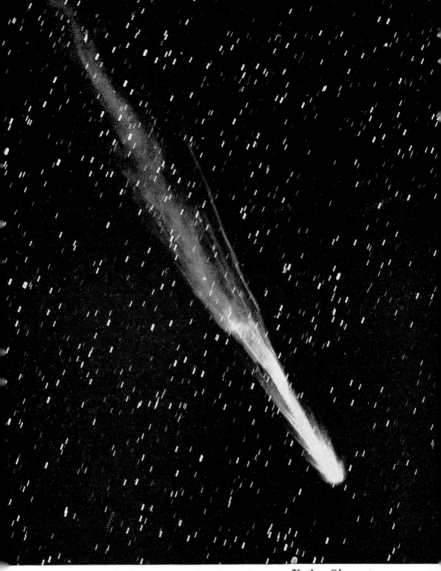

Yerkes Observatory

Morehouse's Comet (the small dashes are caused by the relative displacement of the stars during the time of the exposure, the telescope following the comet exactly).

Mount Wilson and Palomar Observatories

Galaxy M81 in Ursa Major.

Mount Wilson and Palomar Observatories

PHOTOGRAPH OF THE CRAB NEBULA: $\lambda 6300-\lambda 6700$

These are the gases flung out by the supernova of A.D. 1054. They have been traveling outward for nearly nine hundred years at a speed close to 1000 km. per sec., and now the envelope is about five light years across.

Mount Wilson and Palomar Observatories

GALAXY IN ANDROMEDA
The bright dots covering the photograph are foreground stars belonging to our own Galaxy.

Mount Wilson Observatory

GALAXY M33 IN TRIANGULUM

Mount Wilson and Palomar Observatories

GLOBULAR CLUSTER

The direction of this globular cluster lies in the constellation Hercules. It can be seen with a small telescope.

THE ORIGIN OF THE STARS

us from the Sun, and we say that the distance of the Sun is about eight light minutes. I think you will agree that it gives an extremely graphic description of the distances of the stars when I say that light takes about three years to travel to us from even the nearest of them. And when you look at the Milky Way with a small telescope you can see to a distance of more than a thousand light years.

Now we have come to consider the Galaxy, I want to raise a new sort of issue. Where did the Galaxy come from? How are stars born within it? How were our Earth and the planets formed? What is going to be the ultimate fate of the stars? These are samples of the things that I shall be considering from now on. We shall deal in this chapter with the origin of the stars, in a later chapter with the origin of the planets, and finally with the origin of the Galaxy and of the Universe itself.

To make a beginning then, imagine yourself to be looking out across space at the stars of the Milky Way. One of the most important features of the New Cosmology is the realization that this space is not empty at all. Throughout the Milky Way there is a diffuse gas, usually called the interstellar gas. A gas, you will remember, is a swarm of separate atoms and simple molecules. By far the commonest element in the interstellar gas is hydrogen. Hydrogen atoms are ten times more numerous in it as all other atoms and molecules put together. As we shall come increasingly to understand, hydrogen is the basic material out of which the Universe is built.

Now although this gas is so rarefied and although it consists so largely of hydrogen, it isn't all quite pure and it isn't uniformly transparent. For it contains clouds of tiny dust particles, which are a great nuisance to the observational astronomer because they produce a sort of fog that limits his vision whenever he tries to look deep into the Milky Way. Forty years ago it was thought that when we look out at the Milky Way we see the whole of the Galaxy. But we know now that this view is hopelessly wrong. The fog I've just mentioned cuts down our vision so much that, instead of our being able to see the whole of the Galaxy, we see only about a tenth of it.

The Galaxy is a thin disk about 60,000 light years in diameter. It consists of stars and gas. Near its center the disk is very likely much thicker than it is at its edges, where it trails away very gradually. Whereabout in the Galaxy do we lie, our Sun and our planets? The answer to this is: near the edge of the disk. If you want to look toward the center of the Galaxy you should seek out the great star clouds that lie in the constellation of Sagittarius, the Archer. But you will not see the center;

THE ORIGIN OF THE STARS

it is forever hidden from us by the fog we have just discussed. That is to say, you will not see it optically. As you may have heard, certain strange cosmic objects are powerful emitters of radio waves. Radio waves can easily penetrate the fog, whereas light cannot. So if you really want to detect the center of the Galaxy, you should use radio waves and not light.

The interstellar gas is certainly extremely rarefied. On an average over the whole Galaxy a large matchbox full of it would contain only about a hundred atoms. This may be compared with the material in a star, like the Sun, where on the average a matchbox full would contain about a hundred million million million million atoms.

THE BIRTH OF THE STARS

Now I must introduce you to the idea that this immense disk of gas and stars is in motion, that it is turning round in space like a great wheel. How then do the stars move? The main motion of a star is along a path that is roughly a circle with its center at the center of the Galaxy. The Sun and the planets move together as a group around such an orbit. The speed of this motion is in the neighborhood of 500,000 miles an hour. But in spite of this seemingly tremendous speed it nevertheless takes the Sun and its retinue of planets about 200 million years to make a round trip of the Galaxy. At this stage I should like you to reflect on how many ways you are now moving through space. In England you have a speed of about 700 miles an hour round the polar axis of the Earth. You are rushing with the Earth at about 70,000 miles an hour along its pathway round the Sun. There are also some slight wobbles due to the gravitational attraction of the Moon and the other planets. On top of all this you have the huge speed of about 500,000 miles an hour due to your motion around the Galaxy.

The interstellar gas controls the birth of the stars. It is now our business to see how this happens. Astronomers are generally agreed that the Galaxy started its life as a rotating flat disk of gas with no stars in it. There would everywhere be small disturbances in the detailed motions of the various bits of gas, especially near the edge of the disk. To assume a complete absence of such disturbances would be rather like supposing that the flow of water in a whirlpool could be entirely smooth, being devoid of ripples and small eddies.

Now how does a rotating disk of very diffuse gas give birth to compact stars? The first step is a cooling at a number of different, more or less arbitrary places in the disk. The higher pressure of the hotter gas still surround-

THE ORIGIN OF THE STARS

ing these regions then causes compression, thereby forming dense localized clouds. This is confirmed by observation which shows that the interstellar gas is indeed composed of clouds. The distance across an individual cloud usually lies between ten and a hundred light years. So you see that although the clouds are very big when judged by ordinary standards, they are still much smaller than the diameter of the Galaxy. Once clouds have condensed like this, gravitation now exaggerates all the small initial irregularities that they happen to contain. So further condensation takes place in each cloud. At this point it is only necessary to say "and so on," for by repeating the condensation process a sufficient number of times, we must eventually arrive at the particularly dense sort of condensation that we call a star. To sum up the stages —first a whirling disk of gas, then eddies, clouds, condensations, and finally stars.

Granted, then, that pressure and gravitation acting together must lead to the condensation of stars within the rotating galactic disk of gas, let us consider the simplest case of this happening. This is when a star is formed out of a roughly spherical blob of gas. On account of the very diffuse nature of the gas, it is clear that such a blob has to be enormously compressed before a star can be formed out of it. In fact, the blob has to condense to about a millionth of its original diameter. So compared with the gas clouds a star is a body of very small dimensions.

Why does a stellar condensation ever stop contracting? Perhaps I had better clear up this question before we go any further. As a condensation shrinks, its internal temperatures rises, and when this becomes sufficiently high, energy begins to be generated in the interior. This is because a process of atomic transmutation is started up—the process I described in the previous chapter when we were considering how the Sun works. A stage is eventually reached when the energy so generated is adequate to balance the radiation escaping from the surface of the star. Contraction then ceases and the body becomes a normal star like the Sun.

It would be possible to stop at this point and to say that we have explained the origin of the stars. But there are several other questions that trouble the astrophysicist. For instance, we could ask: why is it that all the interstellar gas has not yet been condensed into stars? Or again: why do stars possess some degree of rotation? The answer to this last question is connected with the fact that every cloud of gas and every stellar condensation is in motion around the Galaxy. I cannot explain here exactly why it should be so, but it can be proved that this motion has the effect of generating a rotation in every condensation as it contracts. As I shall be describing in a later chapter, this apparently innocent detail of the condensation process has a profound influence on the formation of planets like the Earth.

STAR GROUPS

Then there is the question: do stars form one one at a time, or do they originate in groups? Once again the more dramatic answer is the correct one. Stars are born in groups. This is pretty obvious from what I have already said, because you wouldn't expect the large-scale eddies and clouds that form within the interstellar gas to end in the making of just one single solitary star. Rather do the clouds break up into a shower of individual condensations, each one becoming a star.

I must emphasize that this whole process is going on all the time. With binoculars you can find lots of these star showers lying along the bright band of the Milky Way. Many hundreds of stars have been born during the last million years within the great gas cloud in the constellation of Orion. Sparkling clusters of recently formed blue stars can be seen within the constellation of Perseus.

The blue stars I've just mentioned are very massive and extremely bright—many of them exceed the Sun in luminosity more than ten thousandfold. These are spendthrift stars with lives spent in a riot of energy. Like butterflies they die almost as soon as they are born. In fact their lives are so short that they never get very far away from the parent cloud of gas. But small stars like the Sun with long lives move far away from their places of birth. By now all traces of the gas cloud in which the Sun was formed have been lost. Even the stars born in the same shower as the Sun have moved apart and are now unidentifiable, like members of a family out of touch with each other.

Stars showers born nowadays contain several hundred members. Spectacular as this may seem, the star groups of the distant past were vastly more populous. At one time stars were born by the million instead of by the hundred. Some of these very old star groups have survived through the aeons. They are known to astronomers as globular clusters—a name prompted by their spherical appearance. One of them in the constellation of Hercules can easily be seen with a small telescope. It is not impossible that one day a globular cluster might pass through the particular bit of the Galaxy where we are located. If this should happen, more than a thousand stars as bright as Sirius could be seen, and there would be a moderate chance that a star belonging to the cluster might

THE ORIGIN OF THE STARS

come close enough to appear as bright as the full Moon.

These large globular clusters move along orbits that lie for the most part quite outside the plane of the Milky Way. In astronomical phraseology, they spend most of their time in the halo of the Galaxy, not in the disk. And this is precisely why they have managed to survive for so long as separate compact groups—the stellar traffic along their orbits has been low. But similar groups in the disk of the Galaxy moved along orbits where the traffic density was high, with the consequence that they were slowly loosened and their constituent stars gradually escaped into the general field. Occasionally, remnants of these old groups can still be found. The cluster Messier 67 is a famous example.

The recently formed smaller groups are disrupted still more easily. Indeed some groups, such as one in the constellation of Perseus, seem to be disrupting spontaneously. Others separate only by passage through dense traffic. The famous group of the Pleiades is not one of the spontaneous disintegrators. But the stellar traffic will inexorably scatter the Pleiades, until in about 500 million years little will remain of this at present spectacular cluster. Here I must remind you that although 500 million years might seem to be a very long time it is in fact much less than the age of the Earth—about 5,000 million years.

THE NATURE OF THE UNIVERSE

From what I have said you will already guess that the general field of stars is really a superposition of many disrupted stellar families—each disintegrating star group contributes a new family. The question arises: can we identify, from the general field, members of particular families? Can we, so to speak, put the stars back again into their original clusters? For the most part the answer is, no. Once the clusters have separated, we mostly find ourselves in the position of the King's men—we can't put them together again. This is so for the Sun's family, as I have already remarked. But very recently O. J. Eggen has indeed succeeded in identifying the members of several very large old families. This important work is still too new for all its implications to be yet known. But it is manifest that the influence of Eggen's discovery on our knowledge of the stars will be profound.

MULTIPLE STARS

Observation shows that most of the stars near the Sun are multiple. That is to say, they consist of two or more stars moving in orbits around each other. The essential point is that the distances between the components of a multiple star are less, often very much less, than the distances which normally separate the stars. In such systems the stellar components stay together and move together around the Galaxy.

A multiple star can be thought of as a very small cluster. Indeed the disintegration of large clusters may be expected to yield many multiple stars. Sometimes two stars within a large cluster become associated together —they come comparatively close to each other and remain together, moving in orbits one around the other. And because of their nearness they tend to remain in association even after the dispersal of the main cluster. In short, the breakup of large clusters yields a multitude of small clusters, sometimes binary with two stellar components, sometimes triple with three components, and so forth.

The remarkable system of Castor is an interesting case. Castor is one of the two bright stars of Gemini, a constellation of the Zodiac. Together with its partner, Pollux, it can be seen on any clear winter night. According to Greek mythology the brothers Castor and Pollux were protectors of ships "when storm winds rage over the ruthless sea." According to modern astronomy, Castor consists of three distinct binary systems, six stars in all. One of these pairs moves around a second in a time of just over 300 years, while the more distant third pair moves around the other two in somewhat more than 10,000 years. Plainly, these pairs are remnants left over from a larger cluster.

A curious point must now be mentioned. The components of binaries are often far closer together than we should expect at first sight. Often the two stars of a pair are separated by a distance less than that of the Earth from the Sun—thousands of times less than the average distance between neighboring stars of the most dense groups. This is an indication that star formation may be a rather gradual process—that stars continue to gain material from the parent gas cloud even after they are first formed; for if they gain material the two components of a binary must come closer together in a remarkable degree. Suppose for instance that a steady condensation process causes the components to increase from half the mass of the Sun to 2.5 times the Sun. Such a fivefold increase causes the distance between the components to decrease to less than one per cent of its initial value. A wide separation could thereby be converted into a rather close separation. And since the increase of mass must be different in different cases, this also explains why binary systems differ quite markedly one from another, particularly so far as the separation of the component stars is concerned.

THE ORIGIN OF THE STARS

STARS IN OTHER GALAXIES

I mentioned at the beginning that there are galaxies other than our own. What of the stars in them? As a broad statement, the stars of other galaxies seem to be quite similar to our own. The main differences lie in the rates at which new stars are being formed. Here we may divide the galaxies into three classes. First, galaxies like the one in Andromeda, the famous Andromeda Nebula. Conditions there seem to be almost exactly the same as they are in the Galaxy. New stars are being born but not at any very rapid rate. The star showers are modest, each apparently comprising a few hundred members. Next, we have galaxies where new stars are scarcely being born at all. These are simply galaxies that do not possess any appreciable quantities of interstellar gas. And of course, because there is no gas, there is nothing to make new stars from. Such galaxies fall into the class known as "the ellipticals." Thirdly, comes the most interesting class, galaxies where star formation is very rapid.

By a lucky chance a member of this third class lies close to hand. The Magellanic Clouds are well-known features of the night sky—only to be seen from the Southern Hemisphere of the Earth, however. These clouds lie near the Galaxy, and may even be appendages of it. In one of them—the Large Cloud—star formation is now proceeding at an intense pace, indeed at the sort of pace that occurred long ago in the Galaxy itself. The exciting thing is that here we have a magnificent picture of what the Galaxy must have looked like at the time of its birth, about 15,000 million years ago.

4 THE FUTURE AND FATE OF THE STARS

IN THIS CHAPTER WE SHALL BE CONCERNED with beginnings and endings. How old is the Galaxy? What is going to be the ultimate fate of the stars and of the Sun in particular? Previously we saw that the Milky Way is all that we can see of a huge disk-shaped cloud of gas and stars that is turning in space like a great wheel. We referred to this cloud as our Galaxy. At one time our Galaxy was a whirling disk of gas with no stars in it. But then, out of the gas, clouds condensed, and then in each cloud further condensations were formed. It is by repetition of this that stars are eventually born, for the stars are simply dense condensations of gas. This is still going on.

Now all this may explain how stars are born, and how the different varieties of star arise. But it does not explain what will eventually happen to the stars, nor does it give us any idea of the age of our Galaxy. These are the questions we must now consider.

Let us begin, as so often we do, with the Earth itself.

The geologists have shown that the Earth must be at least 1,000 million years old, and that throughout this time the Sun must have been shining pretty much as it does now. But geophysicists have done even better than the geologists in getting an idea of the age of the Earth. I cannot explain now exactly how they go to work on the problem, but most of their methods depend on the radioactivity of the uranium present in the rocks of the Earth's crust. Their estimates come out around 4,500 million years. So we can say that our Galaxy must be older than 4,500 million years, because the Galaxy must be older than the Earth.

Next, what information is there about the ages of the stars? The astrophysicist grapples with this question by considering the atomic processes that lead to hydrogen being converted into helium inside normal stars like the Sun. We know the rate at which hydrogen is being consumed in some chosen star. So if we also know how much hydrogen was initially available, it is a fairly straightforward calculation to find how long the supply will last in this star. Every housewife makes similar calculations. Knowing the rate at which you burn coal and how much coal you have got, it is easy enough to see how long it will be before you run out of fuel. The calculations of the astrophysicist are exactly similar to this in principle

THE FUTURE AND FATE OF THE STARS

though more complicated in detail. Hydrogen takes the place of the coal, and the rate at which it is being consumed can be got from the brightness of the star, just as you could get an idea of how much coal you were using from the heat of your fire.

Proceeding in this way we find that in a star as massive as the Sun the hydrogen supply would last for about 15,000 million years, which must therefore be the greatest possible age of the Sun. But much of the Sun's hydrogen is known to be still unconsumed, so we are sure that the Sun cannot yet be as old as this.

But of course the Sun may not have been born at the beginning of the life of the Galaxy, so we must consider stars older than the Sun. Particularly we must look for stars that have just about reached the end of their store of hydrogen. As I explained in an earlier chapter, we can recognize stars in this state; they are the red giants. These stars are the ones that are particularly large in size, and we know that a star cannot have a very large volume unless most of its hydrogen has been used up. Then it only remains to estimate the amount of material present in each such star, and this can be done with considerable accuracy by using the observed brightness. So finally an estimate of the age of each red giant can be obtained. This has been carried out for a number of cases.

The estimate for the stars of the cluster Messier 67 comes out at about 15,000 million years. This is probably a pretty good value for the age of the Galaxy, although it is possible that stars older than Messier 67 may someday be discovered.

This estimate will probably surprise you, as it shows that although the Earth is younger than the oldest stars in our Galaxy it is not enormously younger. In fact the Earth, and probably the Sun too, is about one third of the age of the Galaxy. Also it means that the oldest stars have still only lived long enough for about 75 trips around the Galaxy.

When I was at school, I learned history in such a way as to think a period of a century or two was a very long time. It came as a great shock to realize later that the real history of man must be measured not in centuries but in tens and perhaps in hundreds of thousands of years. But even this is only the briefest tick of the clock compared with the ages of the rocks in your garden and the stars in the sky. What is so important about the time estimates of the astrophysicist is not that the results are staggering almost beyond belief, but that they are reasonably definite and precise, more precise than anything we know about the history of man if you go back more than a few thousand years. We are inescapably faced with the situation

THE FUTURE AND FATE OF THE STARS

that our Galaxy is not a timeless structure, but something that came into being about 15,000 million years ago. How did it come into being? What is the significance of periods of time like this? These are the deeper issues that come out of our present discussion. The answers to them must form a part of our cosmology when later we come to consider the Universe as a whole.

This completes the first part of this chapter, for now we have settled the ages of the stars and the age of our own Galaxy. It remains to deal with the question of the ultimate fate of the stars. Let us begin by considering another exceptional class of star—the supergiants. The supergiants are stars far more massive than the Sun. For definiteness consider a supergiant that contains ten times as much material as the Sun. Such a star would be at least a thousand times brighter than the Sun. The reason for this is that massive stars are extremely prodigal in the rate at which they consume hydrogen. Indeed in such a star the supply of hydrogen can last only for about 100 million years. This is much less even than the age of the Earth. Indeed, the lifetime of a supergiant is so short that it must be a common occurrence in the Galaxy for supergiants entirely to consume their supplies of hydrogen. What happens to them then?

The answer is that the supergiant slowly collapses. This

comes about because it continues to lose energy at its surface. That is to say, the star continues to radiate light and heat into surrounding space, and this loss has to be made good by the star's slowly collapsing inward on itself. In other words the star develops into a collapsed supergiant. As it does so, its central temperature necessarily becomes greater, and the leakage of energy to the surface also becomes greater. So the first effect of the loss of radiant energy at the surface is not to cool off the star but to heat it up. But this is only achieved through the star's living on its capital, through collapsing inward on itself.

How long can such a collapse continue? To answer this question I must now remind you that every star is in rotation. And by a well-known principle in mechanics, as a star collapses its rotation becomes more and more rapid. As it does so the internal forces set up by the rotation become larger and larger. This cannot go on indefinitely. A stage has to be reached at which the rotary forces become comparable with gravity itself. At this stage such a star begins to break up through the power of its own rotation. But this is not the end of the story. We must look a little deeper into the contraction process if we are to understand the different sorts of collapsed stars observed by the astronomer.

So long as the radiation that escapes from the surface

THE FUTURE AND FATE OF THE STARS

of a star like this is the sole cause of the collapse, nothing very violent can happen. The rotary forces increase too slowly for that. What happens is that the star breaks up not in one enormous explosion, but through the steady showering off of material, rather like a gigantic catherine wheel. The steadiness of this process is occasionally punctuated by a sort of spluttering in which a cloud of material, roughly comparable with the Earth in total mass, gets ejected into space with a speed of about 5 million miles an hour. When this happens the hot inner regions of the star become exposed for a while and this leads to a temporary increase in its brightness. Such occurrences are familiar to the astronomer, who refers to them as ordinary novae. But explosions on a far grander scale than this are also observed and are called supernovae. We are constantly being told how terrible the hydrogen bomb is going to be. One hydrogen bomb would be sufficient to wipe out the whole of London. But compared with a supernova a hydrogen bomb is the merest trifle. For a supernova is equal in violence to about a million million million million hydrogen bombs all going off at the same time.

We must now see how supernovae arise. I have mentioned that as a collapsed supergiant shrinks, its internal temperature becomes greater. Reactions between atomic nuclei must become more rapid as the temperature rises. When the temperature has risen about a hundredfold, the conversion of helium into heavy elements like iron becomes very important. At this stage a surprising situation arises. It can be shown that if the collapse proceeds still farther before the rotary forces break up the star, *nuclear reactions must start to absorb energy instead of generating it*. When this stage is reached, the loss of radiation from the surface becomes by comparison quite unimportant, and the star then collapses catastrophically because of a rapid absorption of energy by the nuclear processes and not because of a slow loss of energy at its surface. Instead of being slow and steady, taking hundreds of thousands of years, the collapse becomes swift and catastrophic.

The rotary forces not only grow rapidly but perhaps an even more important process comes into action, a process that operates near the surface of the star. So far we have always been thinking of nuclear processes in the deep inner regions of the stars. These were the regions in which hydrogen was converted to helium—and helium to heavy elements like iron. Near the surface on the other hand much of the original nuclear fuel may still remain. This nuclear fuel is compressed and rapidly heated by the catastrophic collapse. It explodes, literally as a cosmic bomb.

THE FUTURE AND FATE OF THE STARS

To sum up then, the stages in the production of a supernova are these: First a star exhausts in its inner regions the sources of nuclear energy. This it does in stages; hydrogen is transformed to helium, then the helium is converted into heavier elements—ultimately into iron. A collapse arises because of the continual escape of radiation from the surface. As contraction proceeds, rotation becomes more important. The final requirement is that rotation must not break up the star too soon, before the onset of catastrophic collapse. Otherwise the star will simply splutter its way through a long series of ordinary nova eruptions instead of reserving the whole breakup process for one colossal explosion. This is finally triggered by the rapid catastrophic collapse which produces a bomb blast in the nuclear fuel that still remains in the outer regions of the star.

Calculation tells us a good deal about the state of a supernova just before the outburst. The collapse must go on very far before this happens. In spite of the enormous amount of material in the star, it must become about as small in volume as the Earth. It emits hard X-rays from its surface into surrounding space. It is so enormously dense that a matchbox full of material taken from its central regions contains about a hundred tons. Its surface probably rotates with a speed of about 10 million miles an hour. And the time required for its catastrophic outburst is as little as one minute. Indeed if some cosmic jester were to grab hold of the Earth and were to put us near such a body, the whole of the Earth would be entirely crushed and would be spread as a thin scum over the surface of the body. This is not just a piece of whimsical nonsense, because much of the material of the Earth actually was at one time a part of a supernova.

Before we leave these stars there are two or three other points that we ought to consider. Not all the material of a supernova is blown away in the explosion. A dense stellar nucleus, containing perhaps about one-tenth of the original amount of material is left behind. What happens to this remnant? Well, after getting rid of most of its material in the explosion, the remnant is able to cool off. It passes gradually, as it cools, from a blue dwarf to a white dwarf, and it is probably in this way that many of the white dwarfs observed by the astronomer have come into being.

Supernovae have other interests for the astrophysicist. As a recurrent theme in this book we have noticed that hydrogen is the basic material out of which the Universe is built. Helium is common in stars compared with other elements because it is produced in appreciable quantities inside them. The abundances of the rest of the elements are so small that it is natural to ask whether all the material in the Universe started its life as hydrogen. This idea seems very likely to be correct. I think that the other atoms have all been produced within the stars, in particular that the heavy elements, such as iron, have been built up in the supernovae we have just been discussing. The explosions of these stars distribute material in interstellar space, where some of it forms into great clouds of dust particles that we discern with the telescope. It is also likely that some of the material escapes altogether from the Galaxy into surrounding space. I shall mention this again when later we come to consider the origin of the Galaxy itself. Then there are other questions, such as whether the supernovae are powerful emitters of radio waves, and whether they are the main source of the mysterious cosmic rays whose energies are so extraordinary. I shall have to pass these interesting questions with the brief remark that many astronomers believe the answer to them both is yes.

But interesting and exciting as these tremendous explosions are, we have not yet settled the question of the ultimate fate of the stars. So far we have been concerned only with rather massive stars. Now we must consider small stars, stars like the Sun or even smaller.

To deal with the final fate of the Sun let us suppose that the Sun is not going to sweep up further large quantities of interstellar gas. Then the amount of material in the Sun will remain pretty much as it is at present. On this basis the future history of the Sun during the next 10,000 million years or so will follow the lines I have already described when I said that the Sun will grow steadily more luminous as its hydrogen supply is converted into helium, and this will go on until the oceans boil on the Earth. And I then went on to say that as the Sun grills the Earth, it will swell, at first slowly and then with increasing rapidity, until it swallows the inner planets one by one: first Mercury, then Venus, and then possibly the Earth itself.

All this refers to a stage before the Sun's nuclear energy becomes exhausted. Once all nuclear fuels are used up, energy generation will cease, and the Sun will then begin to collapse. Its swollen size will disappear. As it shrinks, the surface will change from the dull red color that must occur in the distended state I have just described. First the surface will warm up to a bright red, then to a white heat, and then to a fierce electric blue. Will the Sun become an exploding star? The answer to this question is no.

When a star like the Sun shrinks to about the size of the Earth, a new form of pressure begins to develop inside. This new pressure is important because it operates

THE FUTURE AND FATE OF THE STARS

without a high temperature's being necessary. When it comes into action, it will allow a star like the Sun to cool off without any further collapse occurring. I cannot explain now exactly why this should be possible, but it is what will happen for small stars. For supergiants, on the other hand, cooling off without an explosion cannot occur, because this new form of pressure is not powerful enough to prevent the collapse of really massive stars.

To end our story of the eventual fate of the solar system: Once the Sun starts cooling off, the escape of radiation from its surface into surrounding space will reduce the temperature in the interior. After about 500 million years the steely blue color of the surface will change to white. The Sun will then be similar to the white dwarfs we have already considered. With the further passage of aeons greater than the present ages of the stars, the surface will cool to a dull red, and then after the elapse of a still greater span of time the light will go out altogether and the Sun will be a black dwarf that moves through space accompanied by its retinue of unlit planets—that is to say, accompanied by those planets that it had not consumed at an earlier stage.

By now we are rather more than halfway through this book. So this is a suitable place to take stock of our position. We have considered the nature of the Earth and planets, of the Sun and the stars, and of our Galaxy. But one outstanding problem has been left over, namely the origin of the Earth and our planets. We shall deal with this in the next chapter. After that we must step outside our Galaxy. We shall find that our Galaxy is but one member of a colossal swarm, and we shall see that this colossal swarm is in a state of organized motion. From there we shall be led to consider the origin of the Universe itself. This then is the program for the rest of the book.

5 THE ORIGIN OF THE EARTH AND THE PLANETS

IN ONE OF SIR ARTHUR CONAN DOYLE'S stories, Holmes remarks, "It was while I was in the carriage . . . that the immense significance of the curried occurred to me." The amazingly slow rotation speed of the Sun gives an equally revealing and subtle clue to the mystery of the origin of the planets. The Sun simply has no business to be rotating as slowly as it actually does. Instead of spinning round once in about 26 days, as it in fact does, our expectation would be that it should rotate in as little as a few hours. This expectation is based on the picture of the origin of stars that we discussed in an earlier chapter. You will remember that stars originate in a rotating disk of interstellar gas. First eddies and clouds are formed and then showers of stars.

You might be tempted to think that the calculations must be wildly wrong. But we can see that things are not too bad because many stars do in fact rotate around in a few hours, just as the calculations show they should. The odd feature is that all the stars for which the theory works out correctly are big stars of considerably more mass than the Sun. Dwarf stars like the Sun all spin slowly. Some process must therefore be systematically at work in dwarf stars to rob them of their rotation. So far so good. Now comes the big question: what process?

Before attempting an answer, I must digress a little to mention an important theorem of dynamics, first proved by Newton. A quantity known as rotational (or angular) momentum can never change in a system that is isolated —unconnected to its surroundings, I mean. Consider the Earth. Although the Earth is not isolated in a strict sense —the Earth's magnetic field reaches out into the interplanetary gas giving a very weak connection with the outside world—it can be taken to be nearly so. Hence our theorem shows that the rotational momentum of the Earth must stay very nearly constant. This means that the spin of the Earth will stay nearly constant. Probably you have often heard the question: what keeps the Earth spinning? Here you have the answer.

Now the Sun and planets together also form a pretty well isolated system, so that the total rotational momentum of the solar system cannot change. How much does the solar system actually possess? The answer forces us immediately to the first step in the solution of

THE ORIGIN OF EARTH AND PLANETS

our main problem, for it turns out that the rotational momentum of the solar system agrees very well with the amount that the Sun ought to have had when it first condensed from the interstellar gas. The essential point is that by far the main contribution to the rotational momentum of the whole solar system now comes from the orbital motion of the planets around the Sun, whereas the calculations that I spoke of earlier led us to expect that the main contribution would come from a rapid spin of the Sun. The rotational momentum exists all right, but it is not located where we expected it to be!

You can easily diagnose what happened. At one time the Sun was indeed spinning quickly just as our calculations indicate. But then rotational momentum was transferred in some way from the Sun to the planets. As this occurred, the spin of the Sun slowed down; while the planets, in order to take up the rotational momentum, moved outward to greater and greater distances from the center. This is a highlight, for we now have before us just the reason why the planets are so distant from the Sun. You will remember that in a model with the Sun represented by a ball the size of a grapefruit the bulk of the planetatry material lies distant a hundred yards or more. The planets lie so far out from the Sun because they have been obliged to store the large rotational momentum that was once contained in a rapid rotation of the primeval Sun.

THE MAGNETIC CLOCK SPRING

We see therefore that our calculations were more incomplete than wrong. They were incomplete because they did not include the possibility that rotational momentum was transferred from the Sun to the planets. For this to be possible a strong connection must have existed between the Sun and the planetary material—otherwise the Sun would be an isolated system, and our theorem would prevent its spin from changing. Here then we have the awkward step: what was the connection?

There is certainly no strong linkage nowadays between Sun and planets. With the planets as compact condensed bodies no strong connection is indeed possible. Evidently then we must go back to a time before the planets condensed, a time when the planetary material was a gas, a gas swirling in a disk around the Sun. This was a time about 5,000 million years ago.

So here is the situation. We have the Sun at the center, with the disk of planetary gas swirling around it. How was it possible to bridge the gap between? In engineering

THE ORIGIN OF EARTH AND PLANETS

language, how was it possible to convey a torque from the Sun to the outlying planetary gas? The only feasible answer is through a magnetic field—a suggestion first made by the Swedish physicist, H. Alfvén. Already more than a century ago, Faraday showed that a magnetic field can be thought of as a collection of lines of force that behave in many ways like elastic strings. So a magnetic field extending outward from the Sun to the planetary material would behave like a vast aggregate of elastic strings. And as the Sun rotated, the strings would become stretched and twisted into a structure rather like a clock spring. By calculation it can be shown that rather a moderate magnetic field would be sufficient to give a clock spring strong enough winding to maintain the required connection from Sun to planets. In short, the torque that conveyed rotational momentum from the Sun to the planetary material was maintained by a magnetic clock spring.

One of the exceedingly important developments of the New Cosmology has been realization that magnetic fields play a vital role in astronomy. Already we have seen that the Earth possesses a magnetic field. So probably does Jupiter. So certainly does the Sun, and so do the stars in general. Even the Galaxy itself has a magnetic field. There may even be magnetic fields pervading all space.

THE NATURE OF THE UNIVERSE

THE GENERAL PICTURE

So far we have said nothing about the origin of the planetary gas and of how it came to be moving in a disk around the Sun. To understand this point, let us go right back to the beginning, to the birth of the Sun itself. The Sun formed as a condensation within the interstellar gas, the condensation starting as a distended blob of gas of enormous dimensions—very much larger than the present scale of the solar system. As time went on, the condensation shrank more and more, and as it did so, rotation prevented the gas from settling into a spherical shape. Condensation to the dimensions of the planetary orbits caused rotation to make the blob more and more flattened at its poles. Condensation to the orbit of Mercury must indeed have produced so marked a distortion that the primitive Sun actually grew a disk at its equator. This was just the disk of planetary gas.

There is no idle speculation here, for astronomical observation provides ample evidence that rapidly rotating stars do in fact grow disks of gas in this fashion. So we see that the genesis of the planetary material was a quite natural and normal stage in the process of star formation.

Next we bring our clock spring into action. The magnetic field existing already inside the solar condensation

THE ORIGIN OF EARTH AND PLANETS

provided a bridge between the primeval Sun and its newly grown disk. The magnetic field became wound into a "clock spring" and rotational momentum began to pass from the Sun to the disk. So as the central solar condensation continued to shrink, it lost more and more rotational momentum, yielding finally a slow spinning star. Meanwhile the planetary material gained rotational momentum. This caused the gas to be pushed farther and farther outward from the center. Starting near the orbit of Mercury, the planetary gas was pushed successively past the orbits of Venus, Earth, and Mars, until ultimately the region of the great planets was reached.

One point before we proceed—can we be quite sure that there was indeed a magnetic field within the solar condensation? Yes, because all stars seem to possess magnetic fields.

THE EARTH

By now we are in a position to bring off a great stroke. First a question: if the planetary gases were pushed far out from the center, to the distances of the great planets, how could the Earth and the other inner planets—Mercury, Venus, and Mars—ever have been formed? Well, certain quite rare materials did not stay in a gaseous form. They condensed out of the gas as solids and liquids in a manner analogous to the formation of raindrops in the clouds of the terrestrial atmosphere. What materials? Those of very low volatility—rocks and metals—in fact exactly the materials of which the inner planets are actually made. And once they had thus condensed from the gas, these materials were no longer subject to the clock spring action of the magnetic field. In short, they were left behind as the main part of the gas swept outward to the orbits of the great planets.

This ties together a whole lot of facts of the case. We know that the atoms present in rocks and metals comprise only about ¼ per cent of the weight of the Sun. The concentration of these atoms in the planetary material must also therefore have been about ¼ per cent. This requires the rock and metal planets—the inner planets, Mercury, Venus, Earth, and Mars—to possess a combined mass no greater than ¼ per cent of the whole planetary system. And this is exactly the case. The inner planets are small, just as they should be.

To restate these points: The innermost planets must be small because they are built out of low volatility materials that are rare. This situation must arise for every star similar to the Sun. Every planetary system will be the same, in that small rock and metal planets will lie on the inside.

THE GREAT PLANETS

The planetary gas would be too hot for much water to condense near the Earth. This is the reason why comparatively little water is to be found on the inner planets, in spite of water molecules being very common. Water must have comprised about one percent of the weight of the gas—considerably more than the ¼ per cent contributed by rock and metal atoms. But the gas must have cooled as it moved away from the Sun so that eventually the water condensed, mainly as huge chunks of ice. Ammonia must also have been common and must similarly have condensed. These indeed were the substances that provided a basis for the building of the great planets.

A contemplation of the terrestrial oceans might raise doubts in your mind about there being very little water on the inner planets! But this is true, comparatively speaking. Our oceans comprise less than one part in 10,000 of the total water content of the whole planetary material. The overwhelming proportion of the water is now to be found in the great planets, not in the small inner planets.

THE NATURE OF THE UNIVERSE

Next, I would like to separate the great planets into two pairs, Jupiter, Saturn and Uranus, Neptune. At the extreme outside of the solar system, gravity was so weak that the very light gases—hydrogen and helium—were able to escape entirely into interstellar space. This is why Uranus and Neptune at the outside are mainly water-ammonia planets with little in the way of hydrogen and helium. Jupiter and Saturn, on the other hand, lying closer in, contain great quantities of hydrogen and helium. And because the original planetary material consisted so overwhelmingly of these light gases (98 per cent or more), this is precisely why Jupiter and Saturn are so much heavier than Uranus and Neptune.

Jupiter and Saturn are therefore the only planets to reflect even approximately the true original composition of the planetary material. The other planets are all abnormal—the inner planets because of the early condensation of rare low volatility materials, the outermost planets because of the entire escape from the solar system of hydrogen and helium.

PLANETS BY THE BILLION

The great majority of stars are slow spinning like the Sun. So it is highly likely that a very similar situation exists in all cases; that planetary systems like our own have developed, even down to details such as small rock and iron planets like the Earth being on the inside, and water-ammonia planets like Uranus and Neptune being on the outside. Not much in our system looks as if it is a matter of sheer accident.

How many slow spinning dwarf stars like the Sun are to be found within the confines of our Galaxy? About 100,000 million. When I said in the first chapter that there might be 1,000 million planets in the Milky Way on which you could live without undue discomfort I was really giving quite a conservative estimate. The precise atmospheric conditions, the precise distances from the central star, would be wrong in many cases for our human needs. But I cannot believe the essentials would be wrong in much more than ninety-nine cases out of a hundred. Even the distance from the central star is really less accidental than you might suppose.

To see this, consider what would happen if the Earth were twice as far away from the Sun. Obviously conditions at the moment would be much too cold. But, as we have seen, the Sun is steadily getting brighter. A right moment would therefore come when the more-distant Earth would warm up to exactly the proper temperature for life as we know it. So you see that distances really do not have to be chosen at all precisely. The rising heat of the Sun would give us our moment of teeming life even if the distance of the Earth from the Sun were varied within quite wide limits. In fact we are indeed living in the age when things are "right" from our point of view—5,000 million years hence the oceans will boil because the Sun will then have become too hot. No life as we understand it will then survive on the Earth.

Which stars are without planets? Almost certainly the rapidly spinning blue giants have no planets, and for a reason that is readily understood. Any attempt to transfer rotational momentum from these stars to an outlying planetary disk is frustrated, not by any lack of a magnetic clock spring, but because the clock spring becomes much too strong. It becomes so strong that the outlying disk gets blown completely away into space. The explosive violence of the process in these cases has been confirmed by the direct observation of stars like Pleione, one of the blue stars of the Pleiades.

LIFE

One next question is: will living creatures arise on every planet where favorable physical conditions occur? No certain answer can be given, but those best qualified to judge the matter, the biologists, seem to think that life would in fact arise wherever conditions were able to support it. Accepting this, we can proceed with greater assurance. The extremely powerful process of natural selection would come into operation and would shape the evolution of life on each of these distant planets. Would creatures arise having some sort of similarity to those on the Earth? The distinguished biologist C. D. Darlington suggests that this is by no means unlikely. To quote Darlington's own words, "There are such very great advantages in walking on two legs, in carrying one's brain in one's head, in having two eyes on the same eminence at a height of five or six feet, that we might as well take quite seriously the possibility of a pseudo man and a pseudo woman with some physical resemblance to ourselves. . . ."

A final question: Will travel between different planetary systems ever be possible? I am sorry to give an unpopular answer, but I believe this to be an uncompromising no. Communication is a different matter. If living creatures at a high technological level exist on planets belonging to any of the nearest thousand stars, it would be feasible to establish communication. A two-way interchange of information would take many centuries to develop. Even so, perhaps we should be starting now?

6 THE EXPANDING UNIVERSE

AT THE RISK OF SEEMING A LITTLE REPETItive I should like to begin this chapter by recalling some of our previous results. One of the things I have been trying to do is to break up our survey of the Universe into distinct parts. We started with the Sun and our system of planets. To get an idea of the size of this system we took a model with the Sun represented by a ball about six inches in diameter. In spite of this enormous reduction of scale we found that our model would still cover the area of a small town. On the same scale the Earth has to be represented by a speck of dust, and the nearest stars are 2,000 miles away. So it is quite unwieldly to use this model to describe the positions of even the closest stars.

Some other means had to be found to get to grips with the distances of the stars in the Milky Way. Choosing light as our measure of distance, we saw that light takes several years to travel to us from near-by stars and that many of the stars in the Milky Way are at a distance of as much as a thousand light years. But the Milky Way is

THE EXPANDING UNIVERSE

only a small bit of a great disk-shaped system of gas and stars that is turning in space like a great wheel. The diameter of the disk is about 60,000 light years. This distance is so colossal that there has only been time for the disk to turn round about seventy-five times since the oldest stars were born—about 15,000 million years ago. And this is in spite of the tremendous speed of about 500,000 miles an hour at which the outer parts of the disk are moving. We also saw that the Sun and our planets lie together near the edge of our Galaxy, as this huge disk is called.

Now we shall go out into the depths of space far beyond the confines of our own Galaxy. Look out at the heavens on a clear night; if you want a really impressive sight do so from a steep mountainside or from a ship at sea. As I have said before, by looking at any part of the sky that is distant from the Milky Way you can see right out of the disk that forms our Galaxy. What lies out there? Not just scattered stars by themselves, but in every direction space is strewn with whole galaxies, each one like our own. Most of these other galaxies—or extragalactic nebulae as astronomers often call them—are too faint to be seen with the naked eye, but vast numbers of them can be observed with a powerful telescope. When I say that these other galaxies are similar to our Galaxy, I do not mean that they are exactly alike. Some are smaller than ours, others are not disk-shaped but nearly spherical in form. The basic similarity is that they are all enormous clouds of gas and stars, each one with anything from 100 million to 100,000 million or so members.

Although most of the other galaxies are somewhat different from ours, it is important to realize that some of them are indeed very like our Galaxy, even so far as details are concerned. By good fortune one of the nearest of them, only about 2,500,000 light years away, seems to be practically a twin of our Galaxy. You can see it for yourself by looking in the constellation of Andromeda. With the naked eye it appears as a vague blur, but with a powerful telescope it shows up as one of the most impressive of all astronomical objects. On a good photograph of it you can easily pick out places where there are great clouds of dust. These clouds are just the sort of thing that in our own Galaxy produces the troublesome fog I mentioned in earlier talks. It is this fog that stops us seeing more than a small bit of our own Galaxy. If you want to get an idea of what our Galaxy would look like if it were seen from outside, the best way is to study this other one in Andromeda. If the truth be known I expect that in many places there living creatures are looking out across space at our Galaxy. They must be seeing much

THE EXPANDING UNIVERSE

the same spectacle as we see when we look at their galaxy.

It would be possible to say a great deal about all these other galaxies: how they are spinning round like our own; how their brightest stars are supergiants, just like those of our Galaxy; and how in those where supergiants are common, wonderful spiral patterns are found. We also find exploding stars in these other galaxies. In particular, supernovae are so brilliant that they show up even though they are very far off. Now the existence of supernovae in other galaxies has implications for our cosmology.

You will remember that in a previous chapter I discussed the way that heavier elements are built up from hydrogen by nuclear processes that take place within the stars. The supernovae were an important component of this picture. The common metals—iron and nickel for example—and the rock-forming elements, magnesium and silicon, owe their origins to the supernovae. The presence of supernovae in other galaxies implies that these materials are present also in other galaxies just as they are in our own. The chemistry of the elements will be much the same everywhere throughout the Universe. Of especial importance, planets will have similar compositions everywhere. Particularly, there will always be small rock and iron planets like the Earth in the inside regions of all planetary systems.

THE NATURE OF THE UNIVERSE

How many of these gigantic galaxies are there? Well, they are strewn through space as far as we can see with the most powerful telescopes. Spaced apart at an average distance of about 3 million light years, they certainly continue out to the fantastic distance of 5,000 million light years. Our telescopes fail to penetrate farther than that, so we cannot be certain that the galaxies extend still deeper into space, but we feel pretty sure that they do. One of the questions we shall have to consider later is what lies beyond the range of our most powerful instruments. But even within the range of observation there are about 1,000 million galaxies. With some 100,000 million planetary systems per galaxy the combined total for the parts of the Universe that we can see comes out at a hundred million million million.

We now come to the important question of where this great swarm of galaxies has come from. Perhaps I should first remind you of what was said when we were discussing the origin of the stars. We saw that in the space between the stars of our Galaxy there is a tenuous gas, the interstellar gas. At one time our Galaxy was a whirling disk of gas with no stars in it. Out of the gas, clouds condensed, and then in each cloud further condensations were formed. This went on until finally stars were born. Stars were formed in the other galaxies in exactly the same way. But we can go further than this and extend the condensation idea to include the origin of the galaxies themselves. Just as the basic step in explaining the origin of the stars is the recognition that a tenuous gas pervades the space within a galaxy, so the basic step in explaining the origin of the galaxies is the recognition that a still more tenuous gas fills the whole of space. It is out of this general background material, as I shall call it, that the galaxies have condensed.

Here now is a question that is important for our cosmology. What is the present density of the background material? The average density is so low that a ten-gallon hat would contain only about one atom. But small as this is, the total amount of the background material exceeds about a hundredfold the combined quantity of material in all the galaxies put together. This may seem surprising, but it is a consequence of the fact that the galaxies occupy only a very small fraction of the whole of space.

The degree to which the background material has to be compressed to form a galaxy is not at all comparable with the tremendous compression necessary to produce a star. This you can see by thinking of a model in which our Galaxy is represented by a 50-cent piece. Then the blob of background material out of which our Galaxy condensed would be only about a yard in diameter. This incidentally is the right way to think about the Universe as a whole. If in your mind's eye you take the average galaxy to be about the size of a bee—a small bee, a honeybee, not a bumblebee—our Galaxy would be roughly represented in shape and size by the 50-cent piece, and the average spacing of the galaxies would be about two yards, and the range of telescopic vision about a mile. So sit back and imagine a swarm of bees spaced about two yards apart and stretching away from you in all directions for a distance of about a mile. Now for each honeybee substitute the vast bulk of a galaxy and you have an idea of the Universe that has been revealed by the large American telescopes.

Next I must introduce the idea that this colossal swarm is not static: it is expanding. There are some people who seem to think that it would be a good idea if it was static. I disagree with the idea, if only because a static universe would be very dull. To show you what I mean by this I should like to point out that the Universe is wound up in several different ways—energy can be got out of the background material in several ways. For one thing the material itself may be hot. Cooling in localized regions then produces compression and condensation in a manner analogous to the onset of star

THE EXPANDING UNIVERSE

formation inside galaxies—except that whole galaxies are now born, not just single stars. Gravitation also supplies energy as galaxies condense. And gravitation again supplies energy during every subsequent condensation of the interstellar gas inside a galaxy. It is because of this energy that a star becomes hot when it is born. A further source of energy lies in the atomic nature of the background material. It seems likely that this was originally pure hydrogen. This does not mean that the background material is now entirely pure hydrogen, because it gets slightly adulterated by some of the material expelled by the exploding supernovae. As a source of energy, hydrogen does not come into operation until high temperatures develop—and this only arises when stars condense. It is this second source of energy that is more familiar and important to us on the Earth.

Now, why would a universe that was static on a large scale, that was not expanding in fact, be uninteresting? Because of the following sequence of events. Even if the Universe were static on a large scale it would not be locally static: that is to say, the background material would condense into galaxies, and after a few thousand million years this process would be completed—no background would be left. Furthermore, the gas out of which the galaxies were initially composed would condense into stars. When this stage was reached hydrogen would be steadily converted into helium. After several hundreds of thousands of millions of years this process would be everywhere completed and all the stars would evolve toward the black dwarfs I mentioned in a previous chapter. So finally the whole Universe would become entirely dead. This would be the running down of the Universe that was described so graphically by Jeans.

One of my main aims will be to explain why we get a different answer to this when we take account of the dynamic nature of the Universe. You might like to know something about the observational evidence that the Universe is indeed in a dynamic state of expansion. Perhaps you've noticed that a whistle from an approaching train has a higher pitch, and from a receding train a lower pitch, than a similar whistle from a stationary train. Light emitted by a moving source has the same property. The pitch of the light is lowered, or as we usually say reddened, if the source is moving away from us. Now we observe that the light from the galaxies is reddened, and the degree of reddening increases proportionately with the distance of a galaxy. The natural explanation of this is that the galaxies are rushing away from each other at enormous speeds, which for the most distant galaxies that we can see with the biggest telescopes become comparable with the speed of light itself.

My nonmathematical friends often tell me that they find it difficult to picture this expansion. Short of using a lot of mathematics I cannot do better than use the analogy of a balloon with a large number of dots marked on its surface. If the balloon is blown up, the distances between the dots increase in the same way as the distances between the galaxies. Here I should give a warning that this analogy must not be taken too strictly. There are several important respects in which it is definitely misleading. For example, the dots on the surface of a balloon would themselves increase in size as the balloon was being blown up. This is not the case for the galaxies, for their internal gravitational fields are sufficiently strong to prevent any such expansion. A further weakness of our analogy is that the surface of an ordinary balloon is two dimensional—that is to say, the points of its surface can be described by two co-ordinates; for example, by latitude and longitude. In the case of the Universe we must think of the surface as possessing a third dimension. This is not as difficult as it may sound. We are all familiar with pictures in perspective—pictures in which artists have represented three-dimensional scenes on two-dimensional canvases. So it is not really a difficult conception to imagine the three dimensions of space as being confined to the surface of a balloon. But then what does the radius of the balloon represent, and what does it mean to say that the balloon is being blown up? The answer to this is that the radius of the balloon is a measure of time, and the passage of time has the effect of blowing up the balloon. This will give you a very rough, but useful, idea of the sort of theory investigated by the mathematician.

The balloon analogy brings out a very important point. It shows we must not imagine that we are situated at the center of the Universe, just because we see all the galaxies to be moving away from us. For, whichever dot you care to choose on the surface of the balloon, you will find that the other dots all move away from it. In other words, whichever galaxy you happen to be in, the other galaxies will appear to be receding from you.

Now let us consider the recession of the galaxies in a little more detail. The greater the distance of a galaxy the faster it is receding. Every time you double the distance you double the speed of recession. The speeds come out as vast beyond all precedent. Near-by galaxies are moving outward at several million miles an hour, whereas the most distant ones that can be seen with our biggest telescopes are receding at over 200 million miles an hour. This leads us to the obvious question: if we could see galaxies lying at even greater distances, would their

THE EXPANDING UNIVERSE

speeds be still vaster? Nobody seriously doubts that this would be so, which gives rise to a very curious situation that I will now describe.

Galaxies lying at only about twice the distance of the farthermost ones that actually can be observed with the new telescope at Palomar would be moving away from us at a speed that equalled light itself. Those at still greater distances would have speeds of recession exceeding that of light. Many people find this extremely puzzling because they have learned from Einstein's special theory of relativity that no material body can have a speed greater than light. This is true enough in the special theory of relativity which refers to a particularly simple system of space and time. But it is not true in Einstein's general theory of relativity, and it is in terms of the general theory that the Universe has to be discussed. The point is rather difficult, but I can do something toward making it a little clearer. The farther a galaxy is away from us the more its distance will increase during the time required by its light to reach us. Indeed, if it is far enough away the light never reaches us at all because its path stretches faster than the light can make progress. This is what is meant by saying that the speed of recession exceeds the velocity of light. Events occurring in a galaxy at such a distance can never be observed at all by anyone inside our Galaxy, no matter how patient the observer and no matter how powerful his telescope. All the events we actually see occurred in galaxies lying close enough for their light to reach us in spite of the expansion of space that's going on. But the struggle of the light against the expansion of space does show itself, as I said before, in the reddening of the light.

As you will easily guess, there must be intermediate cases where a galaxy is at such a distance that, so to speak, the light it emits neither gains ground nor loses it. In this case the path between us and the galaxy stretches at just such a rate as exactly compensates for the velocity of the light. The light gets lost on the way. It is a case, as the Red Queen remarked to Alice, of "taking all the running you can do to keep in the same place." We know fairly accurately how far away a galaxy has to be for this special case to occur. The answer is about 10,000 million light years, which is only about twice as far as the distances that the giant telescope at Palomar is able to penetrate. This means that we are already observing about half as far into space as we can ever hope to do. If we built a telescope a million times as big as the one at Palomar we could scarcely double our present range of vision. So what it amounts to is that owing to the expansion of the Universe we can never observe events

THE EXPANDING UNIVERSE

that happen outside a certain quite definite finite region of space. We refer to this finite region as the observable Universe. The word "observable" here does not mean that we actually observe, but what we could observe if we were equipped with perfect telescopes.

So far we have been entirely concerned with the rich fruits of twentieth-century observational astronomy and in particular with the results achieved by Hubble and his colleagues. We have seen that all space is strewn with galaxies, and we have seen that space itself is continually expanding. Further questions come crowding in: What causes the expansion? Does the expansion mean that as time goes on the observable Universe is becoming less and less occupied by matter? Is space finite or infinite? How old is the Universe? To come to grips with these questions we shall now have to consider new trains of thought. These will lead us to strange conclusions.

First I will consider the older ideas—that is to say, the ideas of the nineteen-twenties and the nineteen-thirties—and then I will go on to offer my own opinion. Broadly speaking, the older ideas fall into two groups. One of them is distinguished by the assumption that the Universe started its life a finite time ago in a single huge explosion. On this supposition the present expansion is a relic of the violence of this explosion. This big bang idea seemed to me to be unsatisfactory even before detailed examination showed that it leads to serious difficulties. For when we look at our own Galaxy there is not the smallest sign that such an explosion ever occurred. This might not be such a cogent argument against the explosion school of thought if our Galaxy had turned out to be much younger than the whole Universe. But this is not so. On the contrary, in some of these theories the Universe comes out to be younger than our astrophysical estimates of the age of our own Galaxy. Another really serious difficulty arises when we try to reconcile the idea of an explosion with the requirement that the galaxies have condensed out of diffuse background material. The two concepts of explosion and condensation are obviously contradictory, and it is easy to show, if you postulate an explosion of sufficient violence to explain the expansion of the Universe, that condensations looking at all like the galaxies could never have been formed.

And so we come to the second group of theories that attempt to explain the expansion of the Universe. These all work by changing with the law of gravitation. The conventional idea that two particles attract each other is only accepted if their distance apart is not too great. At really large distances, so the argument goes, the two particles repel each other instead. On this basis it can be shown that if the density of the background material is sufficiently small, expansion must occur. But once again there is a difficulty in reconciling all this with the requirement that the background material must condense to form the galaxies. For once the law of gravitation has been modified in this way the tendency is for the background material to be torn apart rather than for it to condense into galaxies. Actually there is just one way in which a theory along these line can be built so as to get round this difficulty. This is a theory worked out by Lemaître which was often discussed by Eddington in his popular books.

I should now like to approach more recent ideas by describing what would be the fate of our observable Universe if any of these older theories had turned out to be correct. According to them every receding galaxy will eventually increase its distance from us until it passes beyond the limit of any conceivable astronomical telescope. So if any of the older theories were right we should end in a seemingly empty universe, or at any rate in a universe that was empty apart perhaps from one or two very close galaxies that became attached to our Galaxy as satellites. Nor would this situation take very long to develop. Only about 100,000 million years would be needed to empty the sky of the 1,000 million or so galaxies that we can now observe there.

My own view is very different. Although I think there is no doubt that every galaxy we observe to be receding from us will in about 100,000 million years have passed entirely beyond the limit of vision of an observer in our Galaxy, yet I think that such an observer would still be able to see about the same number of galaxies as we do now. By this I mean that new galaxies will have condensed out of the background material at just about the rate necessary to compensate for those that are being lost as a consequence of their passing beyond our observable Universe. At first sight it might be thought that this could not go on indefinitely because the material forming the background would ultimately become exhausted. The reason why this is not so is that new material appears in space to compensate for the background material that is constantly being condensed into galaxies. This is perhaps the most surprising of all the conceptions described in this book. For I find myself forced to assume that the nature of the Universe requires continuous creation—the perpetual bringing into being of new background material.

The idea that matter is created continuously represents our ultimate goal. It would be wrong to suppose that the idea itself is a new one. I know of references to the con-

THE EXPANDING UNIVERSE

tinuous creation of matter that go back more than 30 years, and I have no doubt that a close inquiry would show that the idea, in its vaguest form, goes back very much further than that. What is new about it is this: it has now been found possible to put a hitherto vague idea in a precise mathematical form. It is only when this has been done that the consequences of any physical idea can be worked out and its scientific value assessed. I should perhaps explain that besides my personal views, which I shall now be putting forward, there are two other lines of thought on this matter. One comes from the German scientist P. Jordan, whose views differ from my own by so wide a gulf that it would be too wide a digression to discuss them. The other line of attack has come from the Cambridge scientists H. Bondi and T. Gold, who, although using quite a different form of argument from the one I adopted, have reached conclusions almost identical with those I am now going to discuss.

The most obvious question to ask about continuous creation is this: where does the created material come from? At one time created atoms do not exist, at a later time they do. The creation arises from a field, which you must think of as generated by the matter that exists already. We are well used to the idea of matter giving rise to a gravitational field. Now we must think of it also giving rise to a creation field. Matter that already exists causes new matter to appear. Matter chases its own tail. This may seem a very strange idea and I agree that it is, but in science it does not matter how strange an idea may seem so long as it works—that is to say, so long as the idea can be expressed in a precise form and so long as its consequences are found to be in agreement with observation. Some people have argued that continuous creation introduces a new assumption into science—and a very startling assumption at that. Now I do not agree that continuous creation is an additional assumption. It is certainly a new hypothesis, but it only replaces a hypothesis that lies concealed in the older theories, which assume, as I have said before, that the whole of the matter in the Universe was created in one big bang at a particular time in the remote past. On scientific grounds this big bang assumption is much the less palatable of the two. For it is an irrational process that cannot be described in scientific terms. Continuous creation, on the other hand, can be represented by mathematical equations whose consequences can be worked out and compared with observation. On philosophical grounds too I cannot see any good reason for preferring the big bang idea. Indeed it seems to me in the philosophical sense to be a distinctly unsatisfactory notion, since it puts the basic assumption out of sight where it can never be challenged by a direct appeal to observation.

Perhaps you may think that the whole question of the creation of the Universe could be avoided in some way. But this is not so. To avoid the issue of creation it would be necessary for all the material of the Universe to be infinitely old, and this it cannot be for a very practical reason. For if this were so, there could be no hydrogen left in the Universe. As I think I demonstrated when I talked about the insides of the stars, hydrogen is being steadily converted into helium throughout the Universe and this conversion is a one-way process—that is to say, hydrogen cannot be produced in any appreciable quantity through the breakdown of the other elements. How comes it then that the Universe consists almost entirely of hydrogen? If matter were infinitely old this would be quite impossible. So we see that the Universe being what it is, the creation issue simply cannot be dodged. And I think that of all the various possibilities that have been suggested, continuous creation is the most satisfactory.

Now what are the consequences of continuous creation? Perhaps the most surprising result of the mathematical theory is that the average density of the background material must stay constant. The new material does not appear in a concentrated form in small localized regions but is spread throughout the whole of space. The average rate of appearance of matter amounts to no more than the creation of one atom in the course of about a year in a volume equal to that of a skyscraper. As you will realize, it would be quite impossible to detect such a rate of creation by direct experiment. But although this seems such a slow rate when judged by ordinary ideas, it is not small when you consider that it is happening everywhere in space. The total rate for the observable Universe alone is about a hundred million, million, million, million, million, tons per second. Do not let this surprise you because, as I have said, the volume of the observable Universe is very large. Indeed I must now make it quite clear that here we have the answer to our question: why does the Universe expand? For it is this creation that drives the Universe. The new material produces a pressure that leads to the steady expansion. But it does much more than that. With continuous creation the apparent contradiction between the expansion of the Universe and the requirement that the background material shall be able to condense into galaxies can be completely overcome. For if the background material is hot, as it certainly would be if it were created in the form of neutrons (which soon decay with a large release of energy into hydrogen atoms), local cooling and compression cause galaxies to condense. Such irregularities constantly arise from the presence of already existing galaxies. So the background material must give a steady supply of new galaxies. Moreover, the created material also supplies unending quantities of atomic energy, since by arranging that newly created material should be composed of hydrogen we explain why, in spite of the fact that hydrogen is being consumed in huge quantities in the stars, the Universe is nevertheless observed to be overwhelmingly composed of it.

We must now leave this extraordinary business of continuous creation for a moment to consider the question of what lies beyond the observable part of the Universe. In the first place you must let me ask: does this question have any meaning? According to the theory it does. Theory requires the galaxies to go on forever, even though we cannot see them. That is to say, the galaxies are expanding out into an infinite space. There is no end to it all. And what is more, apart from the possibility of there being a few freak galaxies, one bit of this infinite space will behave in the same way as any other bit.

The same thing applies to time. You will have noticed that I have used the concepts of space and time as if they could be treated separately. According to the relativity theory this is a dangerous thing to do. But it so

THE NATURE OF THE UNIVERSE

happens that it can be done with impunity in our Universe, although it is easy to imagine other universes where it could not be done. What I mean by this is that a division between space and time can be made and this division can be used throughout the whole of our Universe. This is a very important and special property of our Universe, which I think it is important to take into account in forming the equations that decide the way in which matter is created.

Perhaps you will allow me a short diversion here to answer the question: how does the idea of infinite space fit in with the balloon analogy that I mentioned earlier? Suppose you were blowing up a balloon that could never burst. Then it is clear that if you went on blowing long enough you could make its size greater than anything I cared to specify, greater for instance than a billion billion miles or a billion billion billion miles and so on. This is what is meant by saying that the radius of the balloon tends to infinity. If you are used to thinking in terms of the balloon analogy, this is the case that gives you what we call an infinite space.

Now let us suppose that a film is made from any space position in the Universe. To make the film, let a still picture be taken at each instant of time. This, by the way, is what we are doing in our astronomical observations.

THE EXPANDING UNIVERSE

We are actually taking the picture of the Universe at one instant of time—the present. Next, let all the stills be run together so as to form a continuous film. What would the film look like? Galaxies would be observed to be continually condensing out of the background material. The general expansion of the whole system would be clear, but though the galaxies seemed to be moving away from us there would be a curious sameness about the film. It would be only in the details of each galaxy that changes would be seen. The over-all picture would stay the same because of the compensation whereby the galaxies that were constantly disappearing through the expansion of the Universe were replaced by newly forming galaxies. A casual observer who went to sleep during the showing of the film would find it difficult to see much change when he awoke. How long would our film show go on? It would go on forever.

There is a complement to this result that we can see by running our film backward. Then new galaxies would appear at the outer fringes of our picture as faint objects that come gradually closer to us. For if the film were run backward the Universe would appear to contract. The galaxies would come closer and closer to us until they evaporated before our eyes. First the stars of a galaxy would evaporate back into the gas from which they were formed. Then the gas in the galaxy would evaporate back into the general background from which it had condensed. The background material itself would stay of constant density, not through matter being created, but through matter disappearing. How far could we run our hypothetical film back into the past? Again according to the theory, forever. After we had run backward for about 15,000 million years, our own Galaxy itself would disappear before our eyes. But although important details like this would no doubt be of great interest to us there would again be a general sameness about the whole proceeding. Whether we run the film backward or forward the large-scale features of the Universe remain unchanged.

It is a simple consequence of all this that the total amount of energy that can be observed at any one time must be equal to the amount observed at any other time. This means that energy is conserved. So continuous creation does not lead to nonconservation of energy as one or two critics have suggested. The reverse is the case, for without continuous creation the total energy observed must decrease with time.

We see, therefore, that no large-scale changes in the Universe can be expected to take place in the future. But individual galaxies will change, and you may well want to know what is likely to happen to our Galaxy.

This issue cannot be decided by observation because none of the galaxies that we observe can be much more than 50,000 million years old as yet, and we need to observe much older ones to find out anything about the ultimate fate of a galaxy. The reason why no observable galaxy is appreciably older than this is that a new galaxy condensing close by our own would move away from us and pass out of the observable region of space in only about 50,000 million years. So we have to decide the ultimate fate of our Galaxy again from theory, and this is what theory predicts. It will become steadily more massive as more and more background material gets pulled into it. After about 50,000 million years it is likely that our Galaxy will have succeeded in gathering quite a cloud of gas and satellite bodies. Where this will ultimately lead is difficult to say with any precision. The distant future of the Galaxy is to some extent bound up with an investigation made about thirty years ago by Schwarzschild, who found that very strange things happen when a body grows particularly massive. It becomes difficult, for instance, for light emitted by the body ever to get out into surrounding space. When this stage is reached, further growth is likely to be strongly inhibited. Just what it would then be like to live in our Galaxy I should very much like to know.

To conclude, I should like to stress that as far as the

Universe as a whole is concerned, the essential difference made by the idea of continuous creation of matter is this: without continuous creation the Universe must evolve toward a dead state in which all the matter is condensed into a vast number of dead stars. The details of the way this happens are different in the different theories that have been put forward, but the outcome is always the same. With continuous creation, on the other hand, the Universe has an infinite future in which all its present very large-scale features will be preserved.

7 A PERSONAL VIEW

LOOKING TO THE FUTURE

I COME NOW TO AN ENTIRELY DIFFERENT class of question. With the clear understanding that what I am now going to say has no agreed basis among scientists but represents my own personal views, I shall try to sum up the general philosophic issues that seem to me to come out of our survey of the Universe.

It is my view that man's unguided imagination could never have chanced on such a structure as I have put before you. No literary genius could have invented a story one-hundredth part as fantastic as the sober facts that have been unearthed by astronomical science. You need only compare our inquiry into the nature of the Universe with the tales of such acknowledged masters as Jules Verne and H. G. Wells to see that fact outweighs fiction by an enormous margin. One is naturally led to wonder what the impact of the New Cosmology would have been on a man like Newton, who would have been able to take it in, details and all, in one clean sweep. I think that Newton would have been quite unprepared for any such revelation, and that it would have had a shattering effect on him.

Is it likely than any astonishing new developments are lying in wait for us? Is it possible that the cosmology of 500 years hence will extend as far beyond our present beliefs as our cosmology goes beyond that of Newton? It may surprise you to hear that I doubt whether this will be so. If this should appear presumptuous to you, I think you should consider what I said earlier about the observable region of the Universe. As you will remember, even with a perfect telescope we could penetrate only about twice as far into space as the new telescope at Palomar. This means that there are no new fields to be opened up by the telescopes of the future, and this is a point of no small importance in our cosmology. There will be many advances in the detailed understanding of matters that still baffle us. Of the larger issues I expect a considerable improvement in the theory of the expanding Universe. Continuous creation I expect to play an important role in the theories of the future. Indeed, I expect that much will be learned about continuous creation, especially in its connection with atomic physics. But by and large I think that our present picture will turn out to bear an appreciable resemblance to the cosmologies of the future.

A PERSONAL VIEW

In all this I have assumed that progress will be made in the future. It is quite on the cards that astronomy may go backward, as, for instance, Greek astronomy went backward after the time of Hipparchus. And in saying this I am not think about an atomic war destroying civilization, but about the increasing tendency to rivet scientific inquiry in fetters. Secrecy, nationalism, the Marxist ideology—these are some of the things that are threatening to choke the life out of science. You may possibly think that this might be a good thing, as we have obviously had quite enough of atom bombs, disease-spreading bacteria, and radioactive poisons to last us for a long time. But this is not the way in which it works. What will happen if science declines is that there will be more work, not less, on the comparatively easy problems of destruction. It will be the real science, where the adversary is not man but the Universe itself, that will suffer.

Next we come to a question that everyone, scientist and nonscientist alike, must have asked at some time. What is man's place in the Universe? I should like to make a start on this momentous issue by considering the view of the out-and-out materialists. The appeal of their argument is based on simplicity. The Universe is here, they say, so let us take it for granted. Then the Earth and other planets must arise in the way we have already discussed. On a suitably favored planet like the Earth, life would be very likely to arise, and once it had started, so the argument goes on, only the biological processes of mutation and natural selection are needed to produce living creatures as we know them. Such creatures are no more than ingenious machines that have evolved as strange by-products in an odd corner of the Universe. No important connection exists, so the argument concludes, between these machines and the Universe as a whole, and this explains why all attempts by the machines themselves to find such a connection have failed.

Most people object to this argument for the not very good reason that they do not like to think of themselves as machines. But taking the argument at its face value, I see no point that can actually be disproved, except the claim of simplicity. The outlook of the materialists is not simple; it is really very complicated. The apparent simplicity is only achieved by taking the existence of the Universe for granted. For myself there is a great deal more about the Universe that I should like to know. Why is the Universe as it is and not something else? Why is the Universe here at all? It is true that at present we have no clue to the answers to questions such as these, and it may be that the materialists are right in saying that no meaning can be attached to them. But throughout the history of science, people have been asserting that such and such an issue is inherently beyond the scope of reasoned inquiry, and time after time they have been proved wrong. Two thousands years ago it would have been thought quite impossible to investigate the nature of the Universe to the extent I have been describing it to you in this book. And I dare say that you yourself would have said, not so very long ago, that it was impossible to learn anything about the way the Universe is created. All experience teaches us that no one has yet asked too much.

And now I should like to give some consideration to contemporary religious beliefs. There is a good deal of cosmology in the Bible. My impression of it is that it is a remarkable conception, considering the time when it was written. But I think it can hardly be denied that the cosmology of the ancient Hebrews is only the merest daub compared with the sweeping grandeur of the picture revealed by modern science. This leads me to ask the question: is it in any way reasonable to suppose that it was given to the Hebrews to understand mysteries far deeper than anything we can comprehend, when it is quite clear that they were completely ignorant of many matters that seem commonplace to us? No, it seems to me that religion is but a desperate attempt to find an escape from the truly dreadful situation in which we find ourselves. Here we are in this wholly fantastic Universe with scarcely a clue as to whether our existence has any real significance. No wonder then that many people feel the need for some belief that gives them a sense of security, and no wonder that they become very angry with people like me who say that this security is illusory. But I do not like the situation any better than they do. The difference is that I cannot see how the smallest advantage is to be gained from deceiving myself. We are in rather the situation of a man in a desperate, difficult position on a steep mountain. A materialist is like a man who becomes crag-fast and keeps on shouting: "I'm safe, I'm safe" because he doesn't fall off. The religious person is like a man who goes to the other extreme and rushes up the first route that shows the faintest hope of escape, and who is entirely reckless of the yawning precipices that lie below him.

I will illustrate all this by saying what I think about perhaps the most inscrutable question of all: do our minds have any continued existence after death? To make any progress with this question it is necessary to understand what our minds are. If we knew this with any precision then I have no doubt we should be well on the way to getting a satisfactory answer. My own answer would be that mind is an intricate organization of matter. In so far as the organization can be remembered and reproduced there is no such thing as death. If ordinary atoms of carbon, oxygen, hydrogen, nitrogen, etc., could be fitted together into exactly the structural organization of Homer, or of Titus Oates, then these individuals would come alive again exactly as they were originally. The whole issue therefore turns on whether our particular organization is remembered in some fashion. If it is, there is no death. If it is not, there is complete oblivion.

I should like to discuss a little further the beliefs of the Christians as I see them myself. In their anxiety to avoid the notion that death is the complete end of our existence, they suggest what is to me an equally horrible alternative. If I were given the choice of how long I should like to live with my present physical and mental equipment, I should decide on a good deal more than 70 years. But I doubt whether I should be wise to decide on more than 300 years. Already I am very much aware of my own limitations, and I think that 300 years is as long as I should like to put up with them. Now what the Christians offer me is an eternity of frustration. And it is no good their trying to mitigate the situation by saying that sooner or later my limitations would be removed, because this could not be done without altering me. It strikes me as very curious that the Christians have so little to say about how they propose eternity should be spent.

Perhaps I had better end by saying how I should arrange matters if it were my decision to make. It seems to me that the greatest lesson of adult life is that one's own consciousness is not enough. What one of us would not like to share the consciousness of half a dozen chosen individuals? What writer would not like to share the consciousness of Shakespeare? What musician that of Beethoven or Mozart? What mathematician that of Gauss? What I would choose would be an evolution of life whereby the essence of each of us becomes welded together into some vastly larger and more potent structure. I think such a dynamic evolution would be more in keeping wiith the grandeur of the physical Universe than the static picture offered by formal religion.

What is the chance of such an idea being right? Well, if there is one important result that comes out of our inquiry into the nature of the Universe it is this: when by patient inquiry we learn the answer to any problem, we always find, both as a whole and in detail, that the answer thus revealed is finer in concept and design than anything we could ever have arrived at by a random guess. And this, I believe, will be the same for the deeper issues we have just been discussing. I think that all our present guesses are likely to prove but a very pale shadow of the real thing; and it is on this note that I must now finish. Perhaps the most majestic feature of our whole existence is that while our intelligences are powerful enough to penetrate deeply into the evolution of this quite incredible Universe, we still have not the smallest clue to our own fate.